Basic Electricity
REVISED EDITION

VAN VALKENBURGH,
NOOGER & NEVILLE, INC.

VOL. 2

HAYDEN BOOK COMPANY, INC.
Rochelle Park, New Jersey

Second Printing—1978

The words "COMMON-CORE," with device and without device, are trade marks of Van Valkenburgh, Nooger & Neville, Inc.

ISBN 0-8104-0877-5
Library of Congress Catalog Card Number 76-57841

Copyright © 1954, 1978 by Van Valkenburgh, Nooger & Neville, Inc. All rights reserved including rights under Universal Copyright, International and Pan American Conventions. No part of this book may be reproduced in any form or by any means without written permission of the copyright holder. Requests for permission to reproduce or use materials from this book should be addressed to the publisher.

Printed in the United States of America

Preface to Revision
of
BASIC ELECTRICITY

The COMMON-CORE® Program — *Basic Electricity, Basic Electronics, Basic Synchros and Servomechanisms,* etc. — was designed and developed during the years 1952–1954. On the basis of a job task analysis of a broad spectrum of U.S. Navy electrical/electronics equipment of that era, there was established a "common-core" of prerequisite knowledge and skills. This "common-core" prerequisite was then programed into a teaching/learning system which had as its primary instructional objective the effective training of U.S. Navy electrical/electronics technicians who could understand and apply such understanding in meaningful job problem situations.

Since that time, over 100,000 U.S. Navy technicians have been efficiently trained by this performance-based system. Civilian students and technicians have accounted for hundreds of thousands more. The military and civilian education and training programs in South America, Europe, the Middle East, Asia, Australia, and Africa have also recognized its usefulness with some 12 foreign-language editions presently in print.

Now the foundation of the COMMON-CORE Program, *Basic Electricity,* is being updated and improved. Its equipment job task base has been enlarged to cover the understanding and skills needed for the spectrum of present-day electrical/electronic equipment — modern industrial machines, controls, instrumentation, computers, communications, radar, lasers, etc. Its technological components/circuits/functions base has been revised and broadened to incorporate the generations of development in electrical/electronics technology — namely, from (1) vacuum tubes to (2) transistors and semiconductors to (3) integrated circuits, large scale integration, and microminiaturization.

Educationally, considerable effort has been given to incorporating individualized learning/testing features and techniques within the texts themselves, and in the accompanying interactive student mastery tests.

Notwithstanding the passage of time, the original innovative, basic text-format, system-design elements of the COMMON-CORE Program still stand — this solid framework of proved effectiveness that has been the stimulus for many of the improvements in vocational/technical education.

VAN VALKENBURGH, NOOGER & NEVILLE, INC.

New York, N.Y.

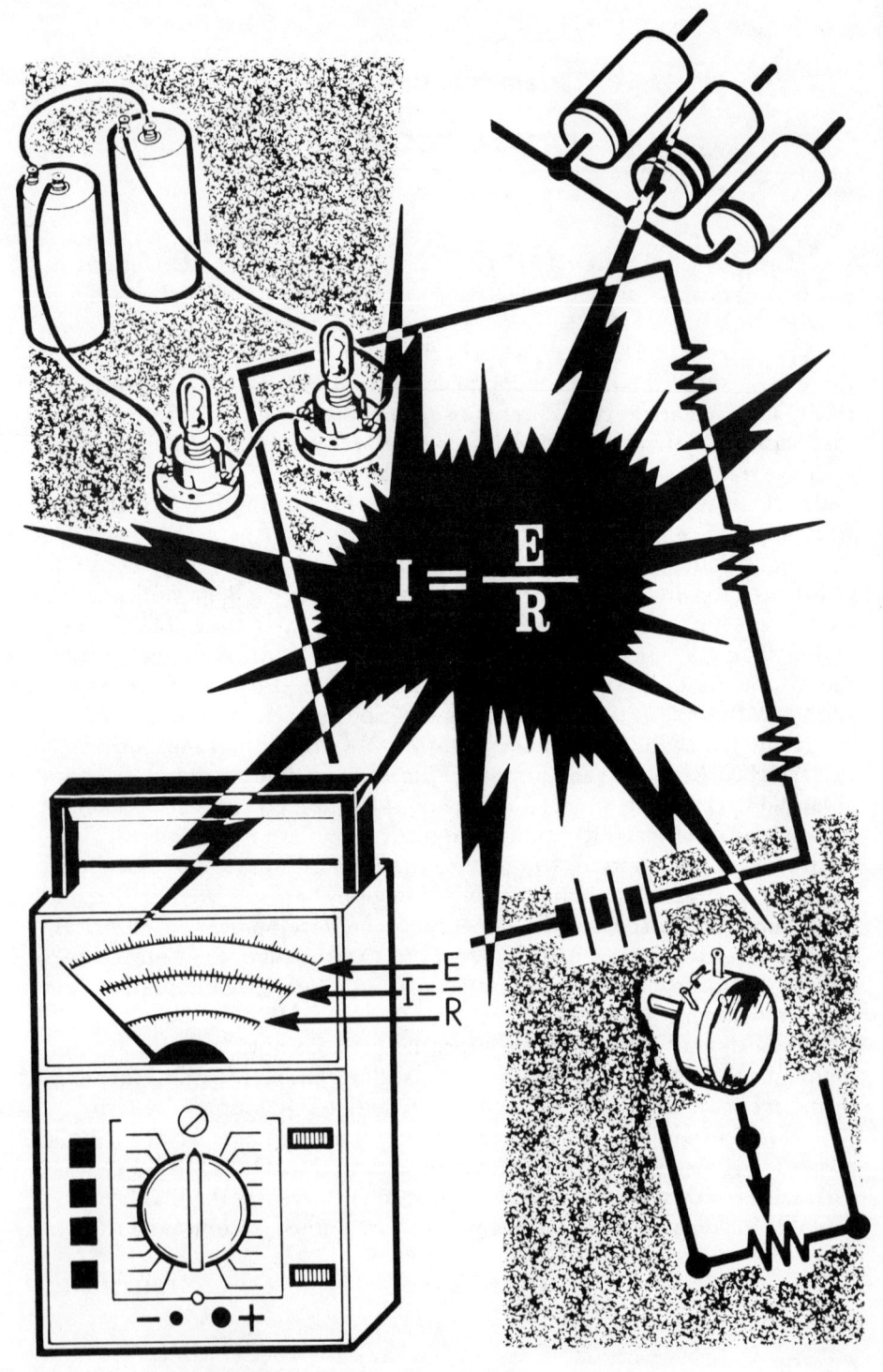

Direct current circuits

CONTENTS

Electric Circuits
What a Circuit Is 2-1
DC and AC Circuits 2-3
The Electric Circuit 2-4
The Load 2-6
Switches 2-7
Simple Circuit Connections 2-8
Review of Electric Circuits 2-9
Self-Test—Review Questions 2-10

Ohm's Law
The Relationship of Voltage, Current, and Resistance 2-11
The Magic Triangle 2-13
Ohm's Law Rules 2-14
Ohm's Law Examples 2-15
Ohm's Law Drill 2-16
Review of Ohm's Law 2-18
Experiment/Application— Ohm's Law 2-19

Resistance
Resistors—Use, Construction, and Properties 2-22
Resistor Tolerance and Values 2-24
Resistor Color Code 2-25
How Resistance Is Measured 2-27
Review of Resistance (Including Material from Volume 1) 2-30
Self-Test—Review Questions (Including Material from Volume 1) . 2-31

DC Series Circuits
The Series Circuit 2-32
Resistance in Series Circuits 2-33
Current Flow in Series Circuits 2-35
Voltage in Series Circuits—Kirchhoff's Second Law 2-36
Ohm's Law in Series Circuits 2-37
Voltage Division in the Series Circuit 2-41
Variable Resistors 2-42
Review of Ohm's Law in Series Circuits 2-45
Self-Test— Review Questions 2-46
Experiment/Application—Open Circuits 2-47
Experiment/Application—Short Circuits 2-50
Experiment/Application—Series Circuit Resistance 2-53
Experiment/Application—Series Circuit Current 2-56

Experiment/Application — Series Circuit Voltage/Kirchhoff's
Second Law 2-57

DC Parallel Circuits
The Parallel Circuit 2-59
Voltage in Parallel Circuits 2-60
Current Flow in Parallel Circuits 2-61
Equal Resistors in Parallel Circuits 2-64
Unequal Resistors in Parallel Circuits 2-65
Experiment/Application — Parallel Circuit Voltage 2-66
Experiment/Application — Parallel Circuit Current 2-67
Experiment/Application — Parallel Circuit Resistance 2-68
Experiment/Application — Parallel Resistances 2-69
Kirchhoff's First Law 2-70
Experiment/Application — Kirchhoff's First Law 2-75
Unequal Resistors in Parallel Circuits (continued) 2-76
Review of Parallel Circuits 2-80
Self-Test — Review Questions 2-81
Applying Ohm's Law in Parallel Circuits 2-82
Solving Unknowns in Parallel Circuits 2-83
Review of Ohm's Law and Parallel Circuits 2-86
Self-Test — Review Questions 2-87
Experiment/Application — Ohm's Law and Parallel Resistances 2-88
Experiment/Application — Ohm's Law and Parallel Circuit Current. 2-90

DC Series-Parallel Circuits
Series-Parallel Circuits 2-91
Resistors in Series-Parallel 2-92
Solving the Bridge Resistor Circuit 2-98
Ohm's Law in Series-Parallel Circuits — Current 2-101
Ohm's Law in Series-Parallel Circuits — Voltage 2-102
Ohm's Law in Series-Parallel Circuits 2-103
Review of Series-Parallel Circuits 2-107
Self-Test — Review Questions 2-108
Experiment/Application — Series-Parallel Connections 2-109
Experiment/Application — Current in Series-Parallel Circuits 2-110
Experiment/Application — Voltage in Series-Parallel Circuits 2-111

Electric Power
What Electric Power Is 2-112
The Power Formula 2-113
Power Rating of Equipment 2-115

Fuses 2-117
Power in Series Circuits 2-119
Power in Parallel Circuits 2-120
Power in Complex Circuits 2-121
Review of Electric Power 2-123
Self Test—Review Questions 2-124
Experiment/Application—The Use of Fuses 2-125
Experiment/Application—How Fuses Protect Equipment 2-126
Experiment/Application—Power in Series Circuits 2-127
Experiment/Application—Power in Parallel Circuits 2-130

Thevenin's and Norton's Theorems
Thevenin's Theorem—Voltage Division
   in the Series Circuit 2-133
Norton's Theorem—Voltage Division
   in the Series Circuit 2-135

Troubleshooting DC Circuits
Troubleshooting DC Circuits—Basic Concepts 2-137
Troubleshooting DC Series Circuits 2-138
Troubleshooting DC Parallel Circuits 2-140
Troubleshooting DC Series-Parallel Circuits 2-142
Drill in Troubleshooting DC Circuits 2-143

Review of DC Electricity
General Review of DC Fundamentals 2-145

Introduction to Alternating Current
Alternating Current 2-148

Answers to Drill Questions 2-149

ELECTRIC CIRCUITS

What a Circuit Is

It is hardly an exaggeration to say that the second half of the twentieth century runs on the flow of electric current. It is, therefore, essential that you should have an accurate picture of what electric current is, and how it behaves in a circuit.

Recall for a moment what you learned about current flow in *Basic Electricity*, Volume 1. You learned that if you connect a length of wire (a conductor) across the positive and negative terminals of a source of electromotive force (emf), say, a battery, the potential difference (voltage) makes the current flow; and also, that electrical energy is needed to keep the current flowing. Additionally, you know that for a battery, the electrical energy is produced from chemical action within the battery.

Many millions of free electrons that have already been separated from the outer orbits of their respective atoms by the heat of room temperature, and which have been wandering aimlessly in all directions through the wire, now come under a *common controlling force*. They are repelled by the *more negative* (or *less positive*) charge which has been set up at one end of the wire, and strongly attracted by the *less negative* (or *more positive*) charge which has been set up at the other end. Their aimless wanderings are converted into a disciplined current flow from more negative to more positive, and electric current flows.

Remember, these electrons are negative charges of electricity and have practically no weight at all. This means that when a potential difference is applied to the wire, they respond to it *immediately*. Similarly, when the potential difference is removed, the electrons stop their disciplined flow in a single direction at once and resume their random wanderings through the conductor material.

IN AN
ELECTRIC CIRCUIT
− +

ELECTRONS ENTER THE WIRE AT THE NEGATIVE TERMINAL

SOURCE of VOLTAGE (EMF)

AND LEAVE THE WIRE AT THE POSITIVE TERMINAL

Any combination of a conductor and a source of electricity connected together to permit electrons to travel around in a continuous stream is called an *electric circuit*.

What a Circuit is (continued)

The conditions required to set up and maintain the flow of an electric current in a circuit are as follows:

1. There must be a *source* of potential difference or voltage to provide the energy which forces electrons to move in a disciplined way in a specific direction.
2. There must be a *continuous (complete) external path* for the electrons to flow from the negative terminal to the positive terminal of the source of voltage.

This external path is usually made up of two parts: the *conductors*, or wires, and the *load* to which the electric power is to be delivered to accomplish some useful purpose or effect. In the illustration below, the load is the lamp.

An electric circuit is thus a completed electrical pathway, consisting not only of a conductor in which the current will flow from negative to positive, but also of a path through a source of potential difference (in this case, the battery) from the positive back to the negative.

A lamp connected across a dry cell battery is an example of a simple electric circuit. Current flows from the negative (−) terminal of the cell, through the lamp (the load), to the positive (+) terminal. The action of the cell is such that it provides a *regenerative* path for the flow of electrons to be maintained.

As long as this electrical pathway remains unbroken at any point, it is a *closed* circuit and current flows. But if the pathway is broken, it becomes an *open* circuit and no current flows.

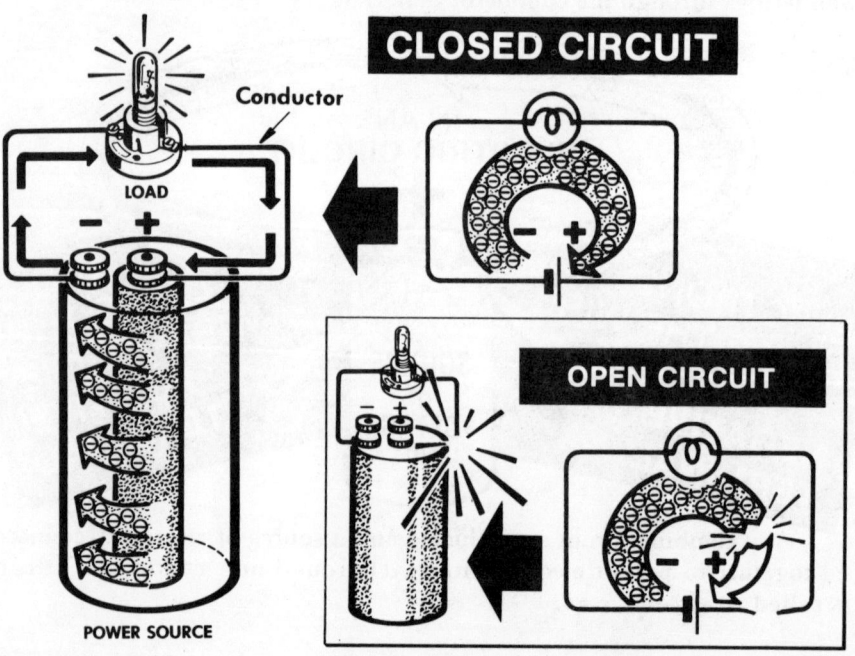

DC and AC Circuits

In electricity we deal with both *direct current* (abbreviated dc) and *alternating current* (abbreviated ac). In dc circuits, the current always flows in the *same* (one) direction. In ac circuits, the direction of current flow *reverses periodically*—in one instant, it will flow in one direction and in the next instant, in the opposite direction. This flow reversal in ac current is usually done regularly so that when we talk about 60-Hz ac power, we mean that the direction of flow reverses 60 times (or cycles) per second.

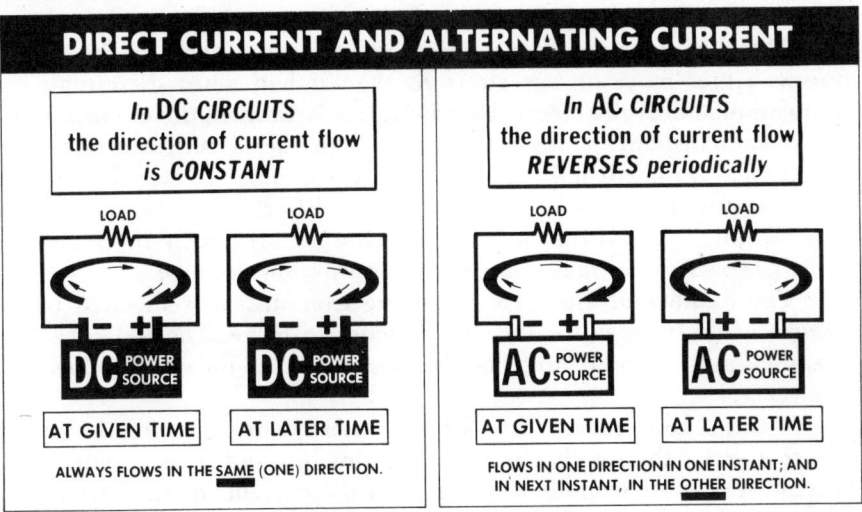

In this volume, we will deal with the function of direct current in circuits containing only resistance (resistive circuits) and we will use Ohm's law and Kirchhoff's laws as the tools for analysis and understanding the relationships of current, voltage and resistance. However, it is important to remember that what you learn here will be *directly applicable* to the ac circuits that you will study in Volumes 3 and 4. By proper interpretation of the concept of current, voltage and resistance, what you study and learn in this volume on dc circuits will be used *again* and *again* for understanding the operation of ac circuits. Therefore, it is *very* important that you completely understand the concepts in dc circuits since they are the foundation for your future understanding of ac circuitry.

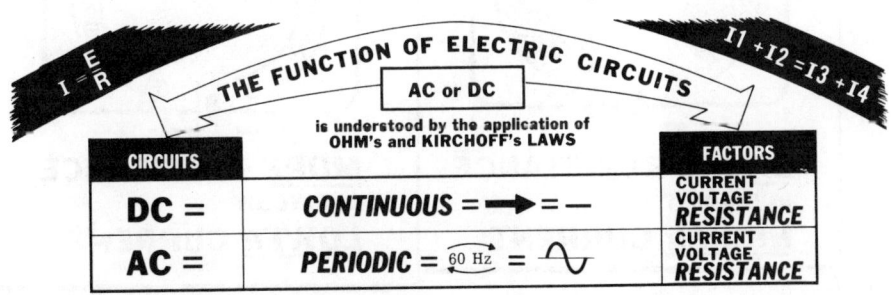

The Electric Circuit

It may help you grasp the concept of an electric current flowing through a closed circuit to imagine that the electrons, which make up the current, form a moving stream which revolves through the completed circuit.

This moving stream of electrons maintains a constant density throughout its entire length. The number of electrons entering the positive terminal of a battery from a wire is always exactly balanced by the number of electrons which the battery forces to move onto its own negative terminal, and, hence, out into the wire.

Thus, at no time does either the conductor wire or the battery possess either more or less electrons than it had when the circuit was first completed. If the circuit loop is suddenly broken, the electron orbiting stream instantly stops *revolving* through the circuit; but both wire and battery will still hold exactly the same number of electrons as they did when the circuit was made. The only difference is that the wire is now holding some of the electrons which were previously in the battery, while the battery has taken an equal number from the wire.

The number of electrons in the electron stream is dictated by the *strength* of the voltage forcing the electrons to move. The *lower* the voltage, the *weaker*—other things being equal—will be the current flow, and vice versa.

When a resistance of any kind is inserted into the circuit loop, it acts to restrict the number of electrons flowing and, hence, reduces the current. You may wonder what restricts the current for the battery and wire circuit that we have been considering. Since all circuits have some resistance, the flow of current is restricted by this resistance.

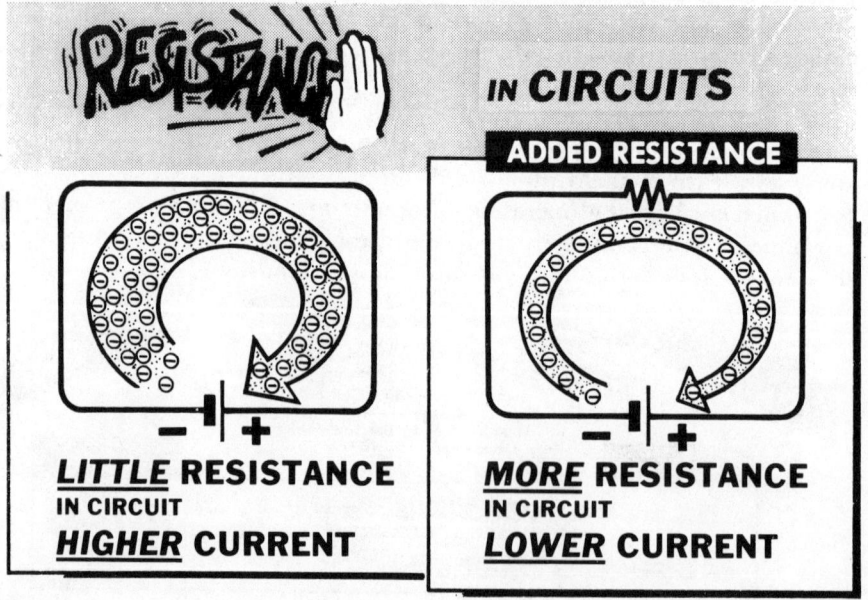

The Electric Circuit (continued)

A closed loop of wire is not always an electric circuit. Only if a source of emf is part of the loop do you have an electric circuit. Current, voltage, and resistance are present in any electric circuit where electrons move around a closed loop. The pathway for current flow is actually the circuit, and its resistance controls the amount of current flow around the circuit.

Direct-current circuits consist of a source of dc voltage, such as batteries, plus the combined resistance of the electrical load connected across this voltage. While working with dc circuits, you will find out how the total load of a circuit can be changed by using various combinations of resistances, and how these combinations of resistances control the circuit current and affect the voltage.

As you will see shortly, there are *two* basic types of circuits: *series circuits* and *parallel circuits*. No matter how complex a circuit you may work with, it can always be broken down into either a series circuit connection or a parallel circuit connection.

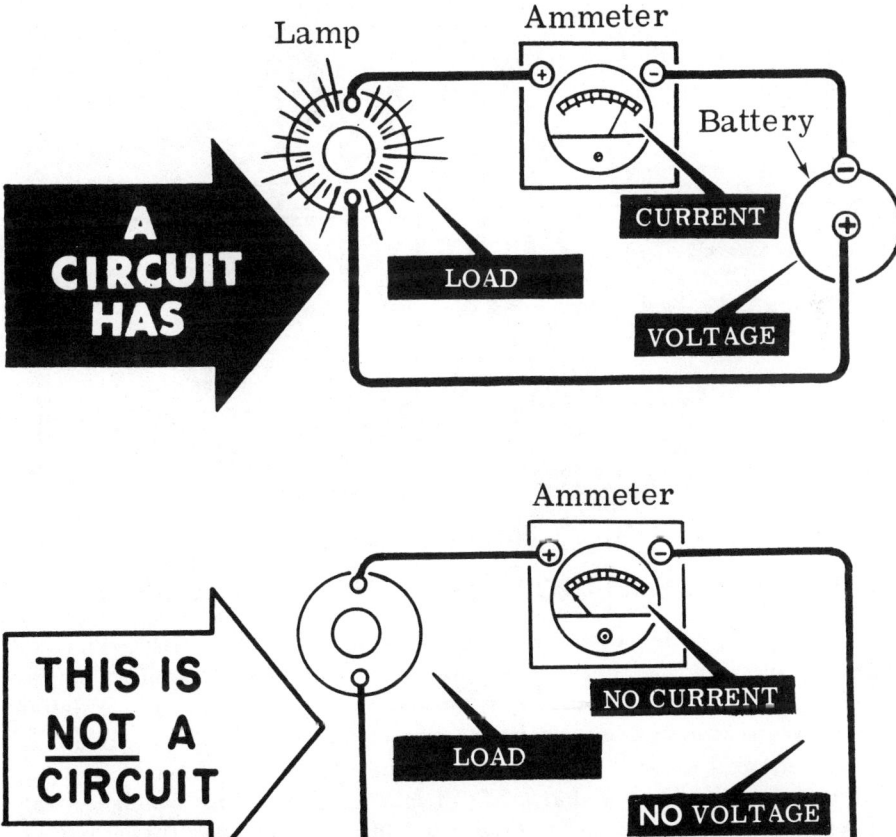

The Load

You learned in Volume 1 that electricity is used to produce pressure (sound), heat, light, chemical action, and magnetism (for mechanical power). In a basic electric circuit, the *device* that transforms the electrical energy from the source of power (emf) into some useful function—such as heat, light, mechanical power, etc.—is called the *load*. The load, besides transforming the electrical energy into one of the foregoing useful purposes, can also be utilized to change or control the *amount* of energy being delivered from the source.

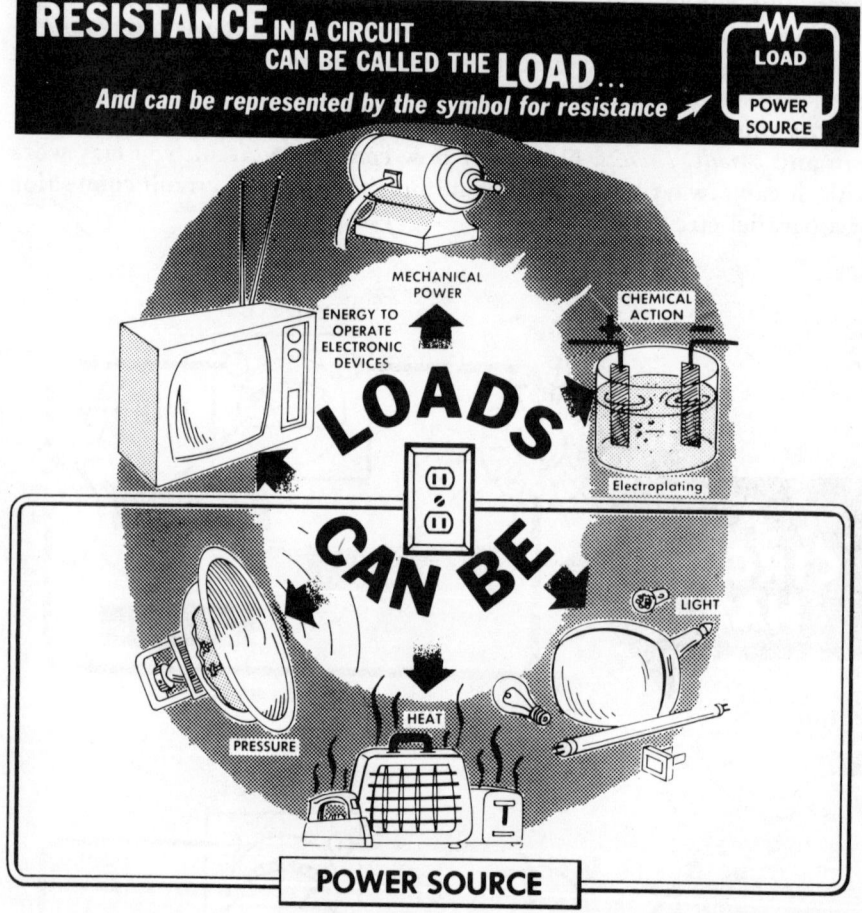

A load can be a motor, a lamp, a telephone, a heater, etc. The amount of electrical energy taken from the source is determined by the type or kind of load. Thus, the term *load* means the *electric power* delivered by the source. For example, when it is stated that the load is decreased or increased, it means that the source is delivering less or more power. Remember that *load* can mean (a) the *device* which utilizes power from the source, and (b) the *power* that is taken from the source.

Switches

A switch is a device used to *open* and *close* a circuit, or part of a circuit, when desired. You have been using switches all your life—in lamps, flashlights, radios, car ignitions, etc. You will meet many other kinds of switches while working with equipment.

You will encounter and use many different switches in your study of electricity. You will also need to know how they are shown symbolically on schematic diagrams. The simplest switch is the *single-pole, single-throw* switch, sometimes abbreviated SPST. More complicated switches can switch many circuits at the same time. These are *multipole, single-throw* switches. A switch that switches *two* circuits is called a *double-pole, single-throw* switch, sometimes abbreviated DPST. In some cases, a circuit is connected to one part of another circuit in one position of a switch, and to another part of the circuit in the other switch position. These are called *double-throw* (DPDT) switches. There are special symbols for these switches in circuit diagrams, as shown in the illustration.

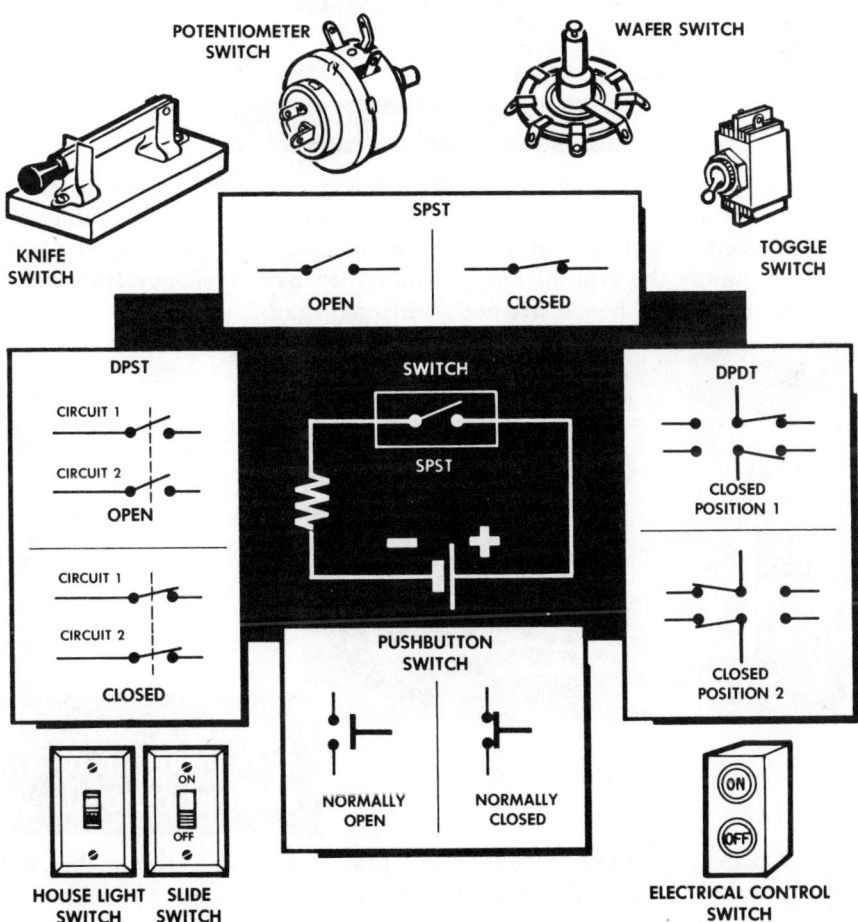

Simple Circuit Connections

Only the loads in the *external* circuit loop, between the terminals of the voltage source, are used to determine the type of circuit. When you have a circuit consisting of only one device, a voltage source, and the connecting wires, it is called a *simple* circuit. For example, a lamp connected directly across the terminals of a dry cell forms a simple circuit. Similarly, if you connect a resistor directly across the terminals of a dry cell, you have a simple circuit since only one device is being used.

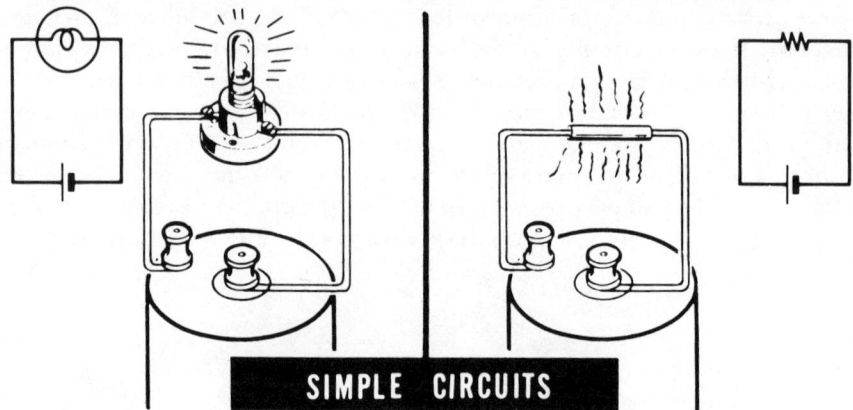

SIMPLE CIRCUITS

Simple circuits may have other devices connected in series with a lamp, but the nature of the circuit does not change unless more than one load is used. A switch and an ammeter inserted in series with the lamp do not change the type of circuit since they have *negligible* (practically no) resistance and, hence, are not additional loads.

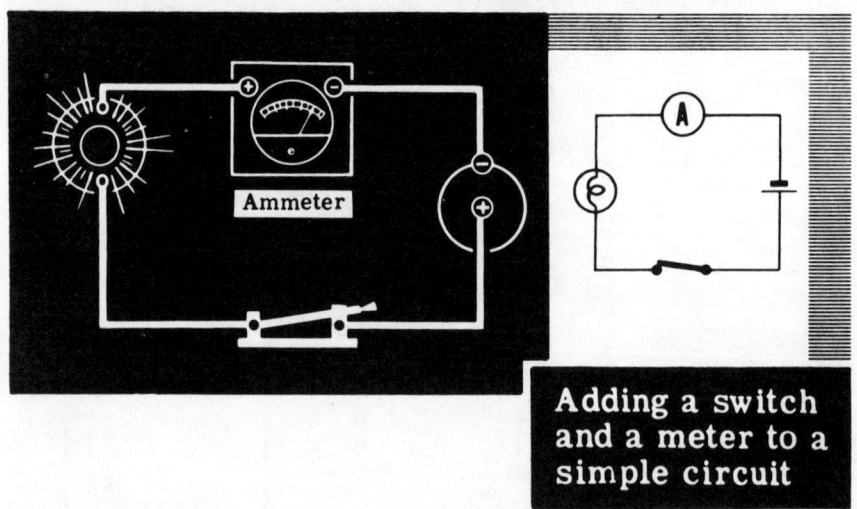

Adding a switch and a meter to a simple circuit

Whenever you use more than one load in the same circuit, they will be connected to form either a *series* or *parallel* circuit, or a combination *series-parallel* circuit.

Review of Electric Circuits

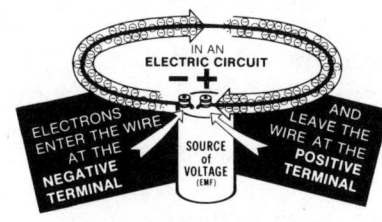

1. ELECTRIC CIRCUIT—A combination of a source of electricity and a conductor that allows electrons to travel in a continuous stream.

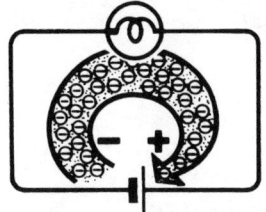

2. CLOSED CIRCUIT—A circuit whose path (loop) is unbroken and current can flow.

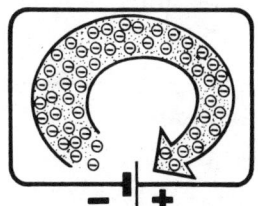

3. RESISTANCE, SMALL—When the resistance is *small*, *large* currents flow.

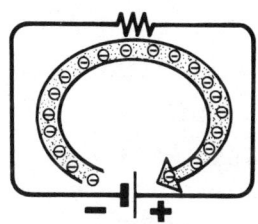

4. RESISTANCE, LARGE—When the resistance is *large*, *small* currents flow.

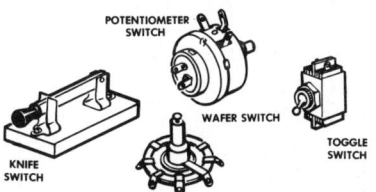

5. SWITCHES—Devices that open and close circuits and, thus, control the flow of electricity.

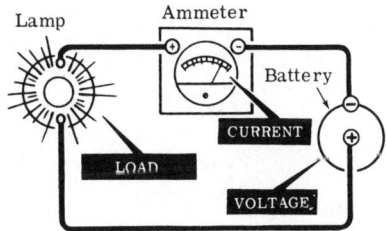

6. LOAD—The device that uses electricity for some function.

Self-Test—Review Questions

1. What are the essential elements in a circuit?
2. Draw a circuit using a battery, conductor, and a lamp.
3. Show a sketch of how you would interrupt the flow of current without a switch.
4. Draw the schematic representation of an SPST switch. Show both open and closed positions.
5. Repeat question 4 for DPST and push-button switches.
6. Draw a circuit using a battery, conductor, and a resistance load with an SPST switch to control current flow.
7. For a constant voltage, the current _____ as the resistance decreases.
8. For a constant voltage, the current _____ as the resistance increases.
9. In the circuit in question 2, which element is the load?
10. Draw a circuit with the battery switchable to two different loads using a DPDT switch.

Learning Objectives—Next Section

Overview—In the next section you will learn about Ohm's Law, one of the most important things that you will use throughout your career in electricity and electronics.

The VOM—The Test Instrument for Ohm's Law

As you will learn, Ohm's Law is concerned with current, voltage, and resistance—$I = E/R$. A test instrument designed to help you obtain the knowns for the above equation is the VOM—the volt/ohm/milliammeter. Measuring any two quantities of the Ohm's Law equation by means of the VOM will provide you with the means for calculating the third—the unknown.

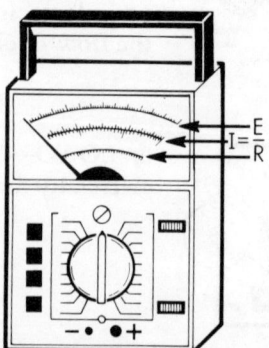

The Relationship of Voltage, Current, and Resistance

You learned in Volume 1 that there exists a fixed relationship of the voltage driving electrons through a circuit to the resistance of that circuit and the rate of current flow through the circuit. It would be wise for you to review these important concepts again.

Given a *constant resistance* in a circuit, you learned that the current flow increases as the voltage applied to the circuit increases. Given a *constant voltage* (emf) applied to the circuit, current flow decreases as the resistance of the circuit increases. You can combine these concepts as follows: *Current flow in a circuit increases as the voltage is increased, and decreases as the resistance is increased.*

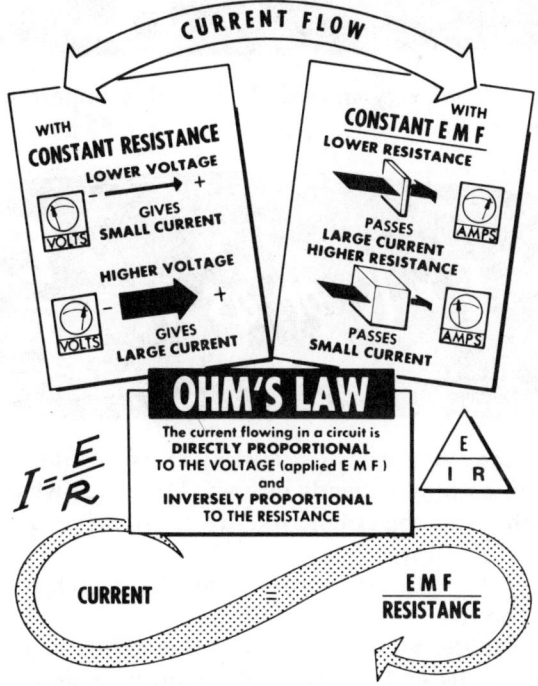

You also learned in Volume 1 that the relationship of voltage, current, and resistance was studied by a German physicist, George Simon Ohm. His statement of this relationship, called *Ohm's Law*, is one of the fundamental laws of physics. You will constantly be using Ohm's Law throughout your work in electricity, as well as later on, whether you intend to study power and light, telephone, electrical machinery, electronics, radar, computers, microwaves, etc.—or, indeed, anything else in which the flow of an electric current is involved.

What, then, does this vital law state? One of the simplest ways of expressing it is given in the illustration above.

It is also possible to express Ohm's Law as a *mathematical equation* (relationship) as further indicated in the illustration above.

The Relationship of Voltage, Current, and Resistance (continued)

In electrical terms (notation), *current* is always represented by the letter "I," *resistance* by the letter "R," and *voltage* by the letter "E." You can, therefore, rewrite the statement of Ohm's Law, at the bottom of the illustration on the last page, as follows:

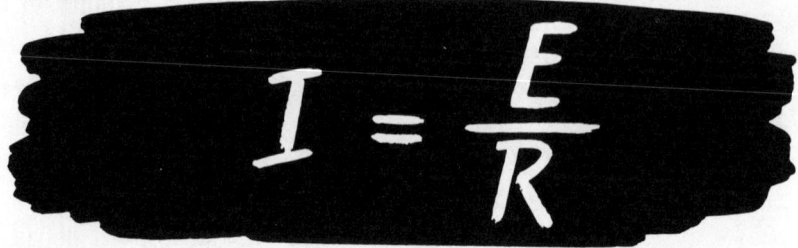

$$I = \frac{E}{R}$$

With the help of very simple algebra, this important equation can also be written as:

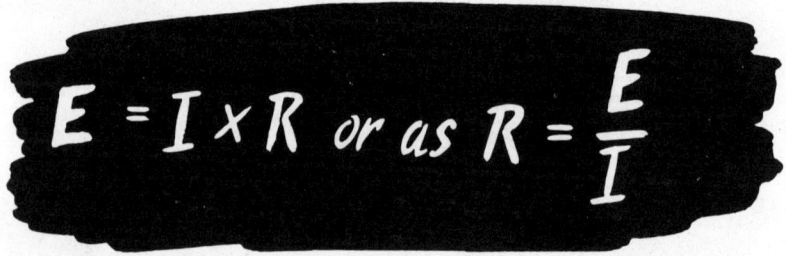

$$E = I \times R \quad \text{or as} \quad R = \frac{E}{I}$$

Which of the three ways (formulas) of expressing Ohm's Law you might choose to employ depends on two things: (a) what *facts* you *know* to start off with about the circuit you are considering, and (b) what *facts* you *need to know* about it.

There is, luckily, an easy way to remember which way or formula to use. Call it, if you like, the *magic triangle*!

Draw a triangle, with a horizontal line across it half-way up from its base. Write the letter E in the small triangle, which has been formed above the line, and write the letters I and R below the line, like so.

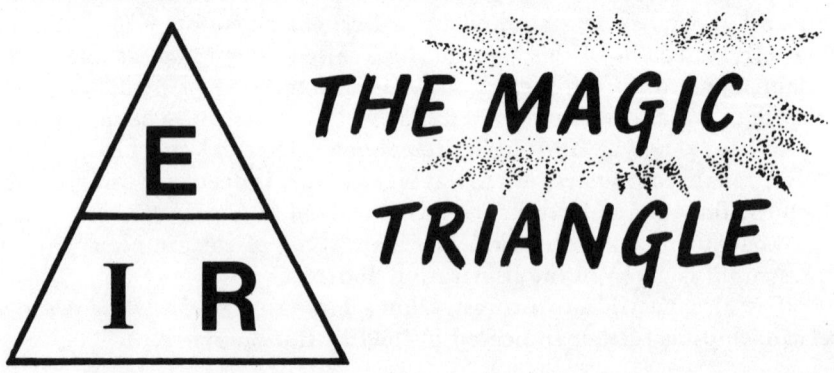

The Magic Triangle

Now consider a circuit in which you know the values of any two of the three factors—voltage, current, and resistance—and want to find out the third. The rule for working the *magic triangle* to give you the correct formula is as follows:

Put your thumb over the letter in the triangle whose value you want to know—and the formula for calculating that value is given by the two remaining letters.

Here is how this useful memory-aid works in practice:

1. You know the values of current and resistance in a circuit, but you lack the means (a voltmeter) to measure the *voltage*. So you draw the magic triangle, put your thumb on the value you want to calculate, which in this case is E—and you are left with the formula you need—I × R.

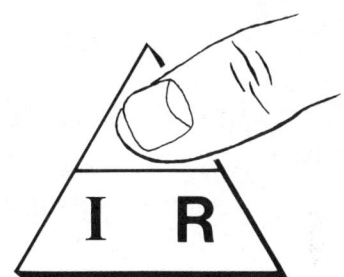

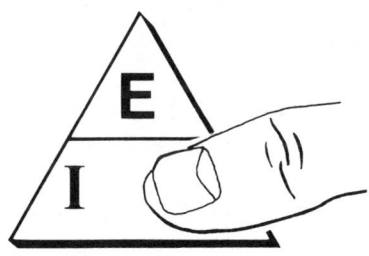

2. You know the values of current and voltage, but in this case you have no ohmmeter to measure the *resistance*. Put your thumb over the letter R and you are left with the formula $\frac{E}{I}$. Substitute the known values for E and I, and your answer is R.

3. The voltage and resistance of a circuit are known to you; but in this case the ammeter you need to measure the *current* is lost or broken. Put your thumb over the symbol I; and read off the formula you need: $\frac{E}{R}$.

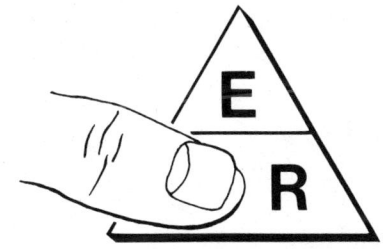

A little thought will show you that the Ohm's Law formula cannot work properly unless all values are expressed *in the correct units of measurement*. The simple rule for this is given on the next page.

Ohm's Law Rules

Ohm's Law will work for you and give you the correct answers to any problem situations which you may try to solve with its help, if you remember that in the Ohm's Law equation, the first rule is that:

CURRENT is *ALWAYS* expressed in **AMPERES**
VOLTAGE is *ALWAYS* expressed in **VOLTS**
RESISTANCE is *ALWAYS* expressed in **OHMS**

Take a circuit in which you have measured the resistance to be 10 ohms, and the current to be 300 milliamperes (mA). Obviously, if you use the Ohm's Law equation blindly and merely write down that $E = I \times R = 10 \times 300 = 3,000$, your answer will be *wrong* by a factor of 1,000.

Instead, you must use the *conversion tables* you learned about in Volume 1, and you must rewrite *all* the factors in the simple equation above in amperes, volts, and ohms. When you do this, you will get:

$$E = I \times R = 10 \times 0.3 = 3\ volts$$

which gives you the correct answer.

There is a second rule which you should apply right from the start whenever you are attempting to solve an Ohm's Law problem involving quantities and values in an electric circuit. The rule is: *Always sketch a rough diagram of the circuit you are considering, before you start making calculations based on the values in the circuit which are already known to you.* This rule, you will find, becomes absolutely essential later on when circuits become more complex.

The sooner you get into the habit of *always* sketching out an Ohm's Law circuit *before* you begin trying to solve it, the better.

Ohm's Law Examples

Ohm's Law, and the correct way to use it, are so central to your electrical training that you should now work through the solution of three simple problems; and then embark on some practice work of your own on Ohm's Law drill.

Example 1

Problem. You have an unknown resistor connected across a battery, and you find by measurement that the voltage across it is 12 volts. You measure the current flowing as 3 amperes. You want to know the resistance of the resistor; but you have no ohmmeter.

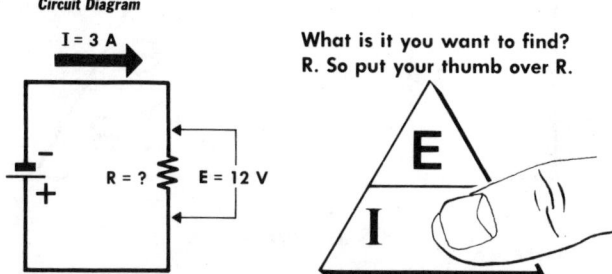

Solution. First draw the circuit diagram, and fill it in with the information you already have. Sketch out the *magic triangle*. The *magic triangle* tells you that $R = \frac{E}{I}$. Into this equation you substitute known values and get

$$R = \frac{12}{3} = 4,$$ which is the value of the resistance in ohms.

Example 2

Problem. What is the voltage across a resistor of 25 ohms when a current of 200 milliamperes is flowing through it?

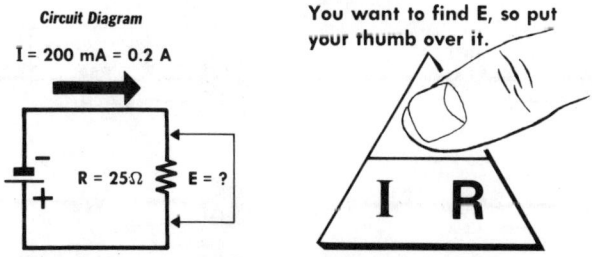

Solution. Sketch the circuit diagram. Draw the *magic triangle*. Convert milliamperes to amperes: $\frac{200}{1,000} = 0.2$ ampere. Then

$$E = IR = 0.2 \times 25 = 5 \text{ volts}$$

Ohm's Law Examples (continued)

Example 3

Problem. A voltage of 60 kilovolts is measured across a resistance of 12 megohms. What current is flowing?

Solution. Recall that a kilovolt is 1,000 volts, so 60 × 1,000 = 60,000 volts. Also, a megohm is a million ohms, so 12 × 1,000,000 = 12 million ohms.

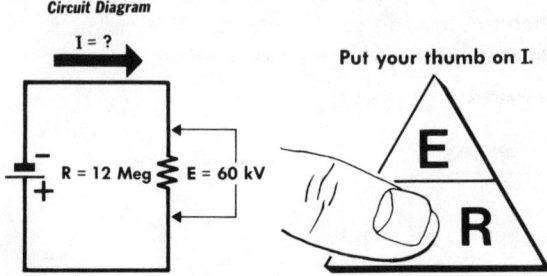

So $I = \dfrac{E}{R} = \dfrac{60,000}{12,000,000} = 0.005$ ampere $= 5$ mA.

(You see that despite the very large voltage applied, the current flowing is a very small one, thanks to the enormous resistance. A circuit with values such as this would seldom be set up in practice; but it is a good one for showing you that Ohm's Law works with *any* values of current, voltage, and resistance—*provided* that you use them correctly.)

Ohm's Law Drill

1. Solve the following:

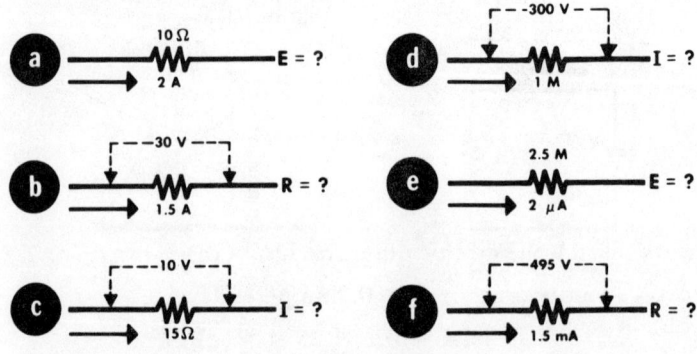

Ohm's Law Drill (continued)

2. You have determined from what is printed on its base, that a bulb from one of the headlights on your car is rated at 12 volts and 4 amperes. What is its resistance?

3. An electromagnet requires a current of 1.5 amperes to make it work properly, and you measured the resistance of its coil to be 24 ohms. What voltage must you apply to make it operate?

4. An electric soldering iron takes 2.5 amperes from a 240-volt supply when it is working. What is the resistance of its element?

5. What is the current through a 68-kilohm resistor when the voltage drop across the resistor is measured at 1.36 volts?

6. What resistance is needed to restrict a current driven by an emf of 10 volts to a flow of only 5 milliamperes?

(Answers to these questions are on page 2-149.)

A Valuable, Practical Tip

It so happens that when you work with practical electronic circuits later on, you will find *two patterns* of values for current, voltage, and resistance tending to recur rather frequently. The values are:

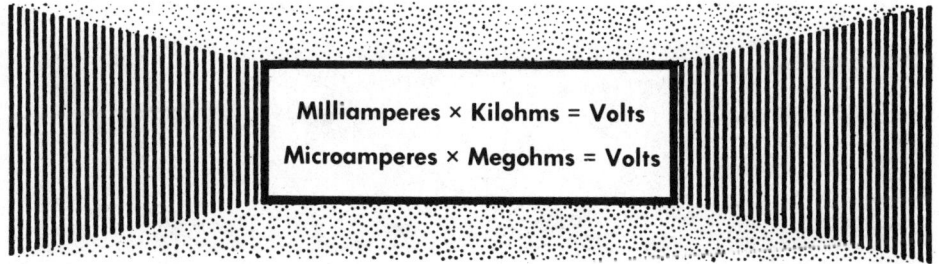

Milliamperes × Kilohms = Volts
Microamperes × Megohms = Volts

If you are going on to study the *Basic Electronics* series after you have finished your work in *Basic Electricity*, you will find it useful to memorize these two *relationships*—*milliamperes* with *kilohms* and *microamperes* with *megohms*—and the fact that they both multiply out into *volts*.

OHM'S LAW

Review of Ohm's Law

Ohm's Law can be stated as a mathematical tool which is of the greatest use in determining an *unknown* factor of current, voltage, or resistance in an electric circuit in which the other two factors are *known*. It can, therefore, be used to take the place of an ammeter, voltmeter, or ohmmeter, respectively, when you are trying to resolve a circuit value in which you already know the two other values.

1. Ohm's Law can be stated in several ways. One of the most useful is this: *The current flowing in a circuit is directly proportional to the voltage applied to the circuit, and inversely proportional to the resistance of the circuit.*

2. This can be stated as an equation:

3. This equation, in symbols, reads: $I = \dfrac{E}{R}$

4. Remember the use of the *magic triangle* in helping you to decide the formula to use. (E on top, I and R below the line. Put your thumb on the quantity you do *not* know, and read off the formula for finding it.)

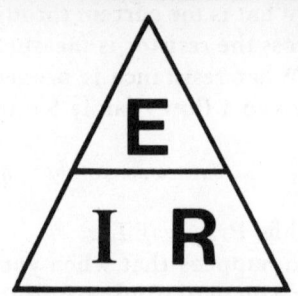

5. Remember that none of the forms in which the Ohm's Law formula can be expressed will work for you unless you keep in mind that:

CURRENT is **ALWAYS** expressed in **AMPERES**
VOLTAGE is **ALWAYS** expressed in **VOLTS**
RESISTANCE is **ALWAYS** expressed in **OHMS**

The Ohm's Law formula is a basic tool for all who work with electric circuits in any shape or form. And, like any tool, it becomes easier to use with practice; and the more often you use it, the more skilled you will find yourself becoming in its application.

Experiment/Application—Ohm's Law

To show how you can apply your understanding of Ohm's Law to find the resistance needed, suppose you connect four dry cells to form a 6-volt battery. Then, if you choose desired values of current such as 0.3, 0.6, and 1 ampere, you can determine the resistance, using Ohm's Law, which will give these currents when connected across the 6-volt battery. The voltage—6 volts—is divided by the desired currents—0.3, 0.6, and 1 ampere—giving required resistances of 20, 10, and 6 ohms, respectively. To check these values, connect two 3-ohm resistors in series to form a 6-ohm resistance, and connect it in series with an ammeter across the 6-volt battery. You will see that the resulting current is approximately 1 ampere. By adding more resistors in series to form 10- and 20-ohm resistances, you can show that these resistance values also result in the desired currents.

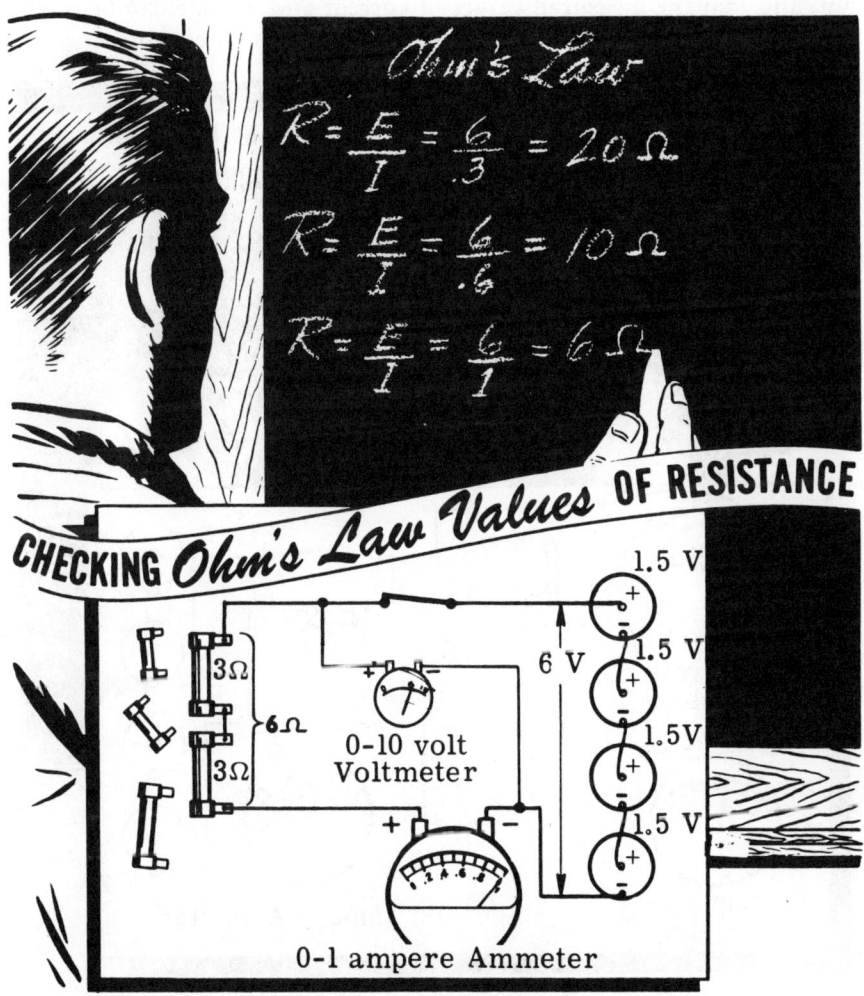

Experiment/Application—Ohm's Law (continued)

Current and voltage may also be used to find the value of a resistance in a circuit when the resistance is unknown. To see for yourself this use of Ohm's Law, suppose you connect two resistors, having no resistance marking, to form a series circuit across a 6-volt battery with an ammeter connected to measure the current flow. When the voltage across each resistor is read, you will see that these two voltages, when added, equal the battery voltage. By dividing the voltages across the resistors by the circuit current, you can obtain the resistance value of the resistors.

To show that your answers are correct, the resistances can be measured with an ohmmeter; it will be found that the values obtained by Ohm's Law equal those indicated on the ohmmeter. As several such problems are worked out, you will see that the rated current and voltage can be used to find the value of the resistance needed in a particular circuit, and that the measured values of current and voltage can be used to find the value of an unknown resistance in a particular circuit.

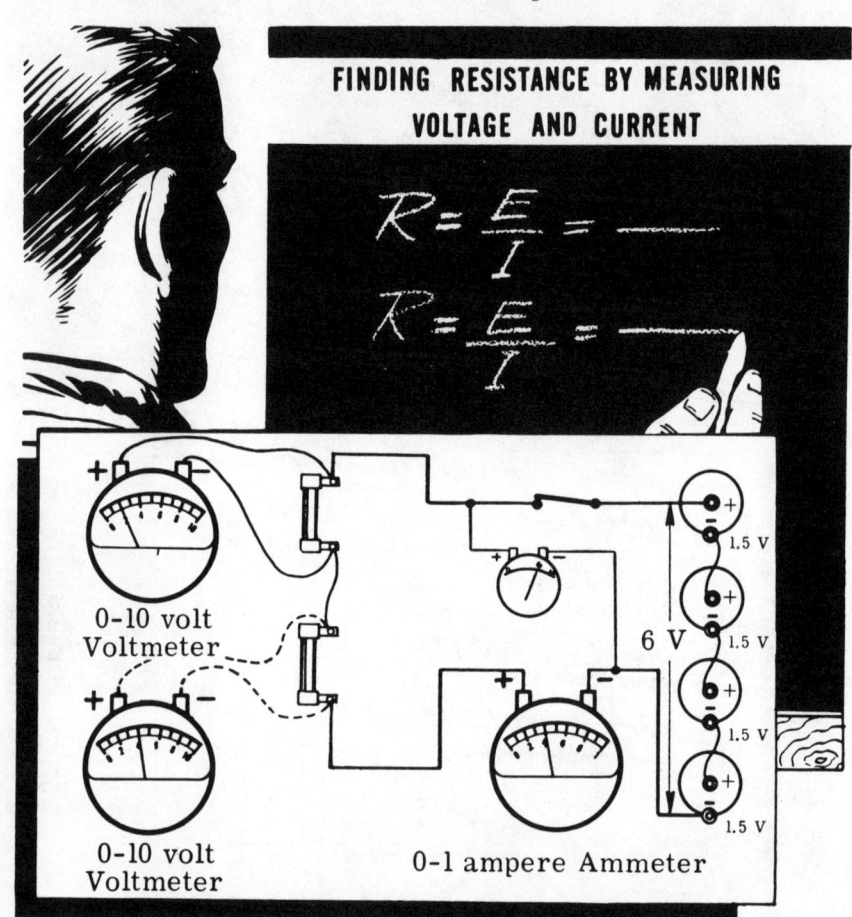

OHM'S LAW

Experiment/Application—Ohm's Law (continued)

You will now see how you can use Ohm's Law to find the voltage required to give the correct current flow through a known resistance. Using a 10-ohm resistance consisting of two 2-ohm resistors and two 3-ohm resistors in series, you can determine the voltages needed to obtain 0.3, 0.6, and 0.9 ampere of current flow by multiplying 10 ohms by each current in turn. The voltage values obtained are 3, 6, and 9 volts, respectively.

To check these values, the 10-ohm resistance can be connected in series with an ammeter across cells connected to give these voltages. With the 3-volt battery of cells, you will see that the current is 0.3 ampere; with the 6-volt battery, it is 0.6 ampere; and with the 9-volt battery, it is 0.9 ampere—showing that the Ohm's Law values are correct.

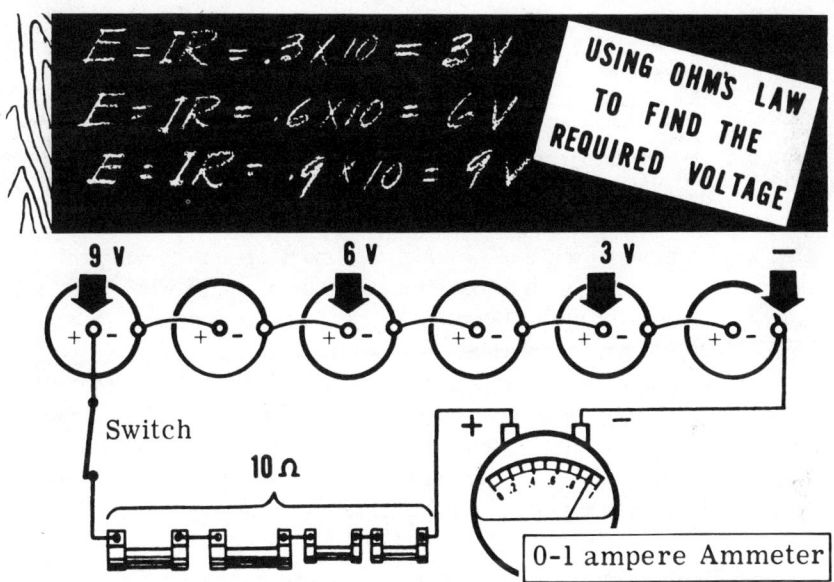

To confirm that Ohm's Law can be used to find the current in a circuit, you can connect six 3-ohm resistors in series across the terminals of a 9-volt battery of dry cells. Using Ohm's Law, determine the circuit current: $I = E/R = 9/18 = 0.5$ ampere. Now break the circuit and insert a 1.0-ampere ammeter in series with the resistors. You will see that the meter reading is 0.5 ampere—exactly the value of current determined by your Ohm's Law calculation.

Learning Objectives—Next Section

Overview—Now that you have learned Ohm's Law, and some of its applications, you can learn about the construction of resistors, some of their other properties, and how resistance is measured.

Resistors—Use, Construction, and Properties

There is a certain amount of resistance in all of the electrical equipment which you use. However, sometimes this resistance is not enough to control the flow of current to the extent desired. When *additional* control is required—for example, when starting a motor—resistance is purposely added to that of the equipment. In the circuit shown, a switch and a current-limiting resistor are used to control the flow of current through the motor. When starting the motor, the switch is kept open and the *resistance* is thereby *added* into the circuit to control the flow of current. After the motor has started, the switch is then closed in order to *bypass* the current-limiting resistor. Before you continue your study of circuits, you need to know more about resistance and resistors.

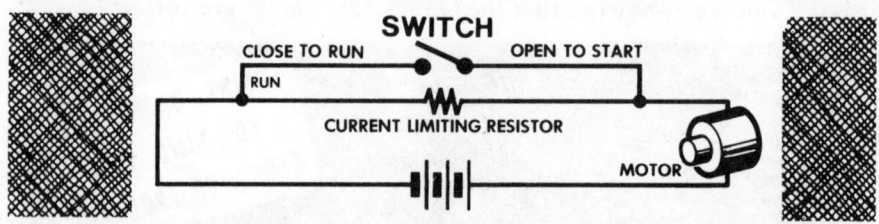

You will use a wide variety of resistors, some of which have a *fixed* value and others which are *variable*. Resistors are made of special resistance wire, graphite (carbon) composition, or of metal film. Wire-wound resistors are usually used to control large currents, while carbon resistors control currents which are relatively small.

Vitreous enameled wire-wound resistors are constructed by winding resistance wire on a porcelain base, attaching the wire ends to metal terminals, and coating the wire and base with powdered glass and baked enamel to protect the wire and conduct heat away from it. Fixed wire-wound resistors with a coating other than vitreous enamel are also used.

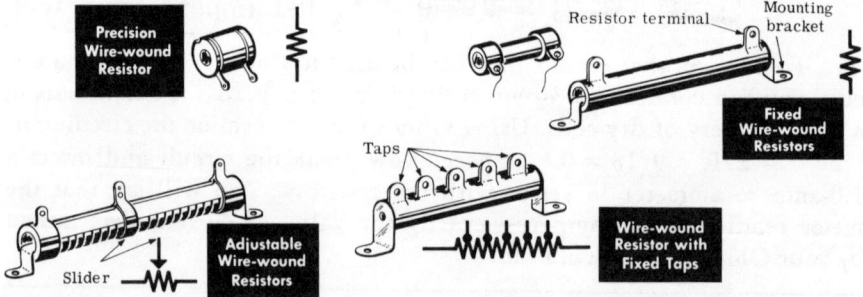

Wire-wound resistors may have fixed taps, which can be used to change the resistance value in steps, or sliders, which can be adjusted to change the resistance of any fraction of the total resistance.

Precision wound resistors of Manganin wire (a special wire that does not change resistance very much with temperature) are used where the resistance value must be very accurate, such as in test instruments.

Resistors—Use, Construction, and Properties (continued)

Generally, carbon resistors are used for low current applications. They are made from a rod of compressed graphite (carbon) that is mixed with clay and binders. By varying the amount of each component, it is possible to vary the resistance values obtained over a very wide range. Two lead wires called *pigtails* are attached to the end of the resistance rod, and the rod is embedded in a ceramic or plastic covering, leaving the pigtails protruding from the ends.

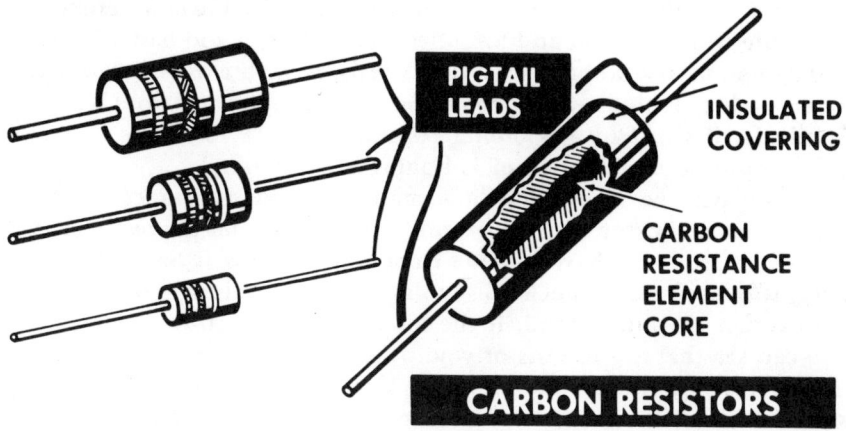

CARBON RESISTORS

Occasionally, you will find a type of resistor called a *deposited film resistor* used for special applications. These resistors are made by depositing a thin film of resistance metal or carbon on a ceramic core and then coating the resistor with either a ceramic or enamel protective coating. In many cases, you will find that these resistors have radial leads; that is, the leads come off at right angles to the body of the resistor. Also, in some cases the deposited film is laid down on the core as a spiral, similar to winding a wire around the tube, in order to increase the length of the resistance element without making the resistor too long.

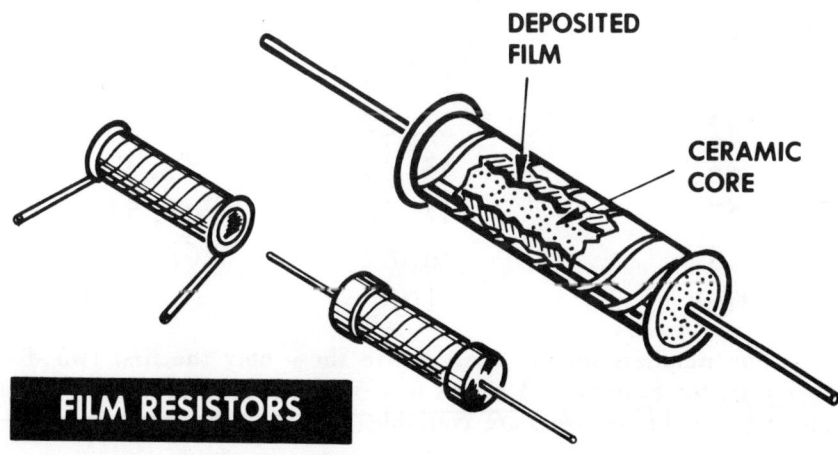

FILM RESISTORS

Resistor Tolerance and Values

Before you go on to the Color Code for resistors, you will need to find out something about resistor tolerances and something about the preferred values of resistance that you will find in circuits. It is very difficult to make a resistor to an *exact* value. Fortunately, in most cases, an approximate value of resistance will do very well. While special resistors may have tolerances of as little as 1%, 0.1%, or even 0.01%, most resistors that you will see have much greater tolerances. Large wire-wound resistors usually have tolerances of 10% or 5%. Carbon resistors are available in 20%, 10%, and 5% tolerances. Thus, if you had a 10-kilohm (10,000-ohm, also abbreviated 10-K) resistor with a 20% tolerance, the actual value of the resistor could be anywhere from 8 to 12 kilohms. Similarly, if you had a 330-ohm resistor with 5% tolerance, the actual value could be anywhere from 314 ohms to 347 ohms.

You may be wondering how many different resistance values one can get for a resistor. As it turns out, this depends on the *tolerance*. Since a 10-K resistor can have a value from 8 to 12 K if it has a 20% tolerance, it doesn't make much sense to make a 9-K resistor with a 20% tolerance. On the other hand, if the tolerance of the 10-K resistor is 5%, you can see that a 9-K resistor would not overlap the tolerance values of a 10-K resistor, and would be useful if such tight tolerances were necessary. Considerations such as this have led to the establishment of a set of preferred values of resistance in each tolerance where the highest tolerance of one value is about equal to the lowest tolerance of the next highest value. These preferred resistance values are shown in the table below. Later, when you learn about power ratings, you will find that resistors are available in different power ratings as well.

PREFERRED CARBON RESISTOR VALUES

20% Tolerance	10% Tolerance	5% Tolerance
10	10, 12	10, 11, 12, 13
15	15, 18	15, 16, 18, 20
22	22, 27	22, 24, 27, 30
33	33, 39	33, 36, 39, 43
47	47, 56	47, 51, 56, 62
68	68, 82	68, 75, 82, 91
100	100	100

The numbers on the chart above show only the first two digits; therefore, for example, 33 means that 3.3, 330, 3.3-kilohm, 330-kilohm, and 3.3-megohm resistors are available.

Resistor Color Code

You can find the resistance value of any resistor by using an ohmmeter; but in some cases, it is easier to find the value of a resistor by its marking. Most wire-wound resistors have the resistance value printed in ohms on the body of the resistor. If they are not marked in this manner, you must use an ohmmeter. Precision wire-wound resistors usually have all of the data printed directly on the resistor body, often including such information as tolerance, temperature characteristics, and exact resistance value. Carbon resistors usually do not have the data on their characteristics marked directly on them; instead they have a *color code* by which they can be identified. The reason for this is that some carbon resistors are so small that the written data would be impossible to read. In addition, carbon resistors are often mounted so that it would be very difficult to read printed values.

Carbon resistors are of two types, radial and axial, which differ only in the way in which the wire leads are connected to the body of the resistor. Both employ the same color code, but the colors are painted in a different manner on each type. Radial-lead resistors are not found in modern equipment, although they were widely used in the past.

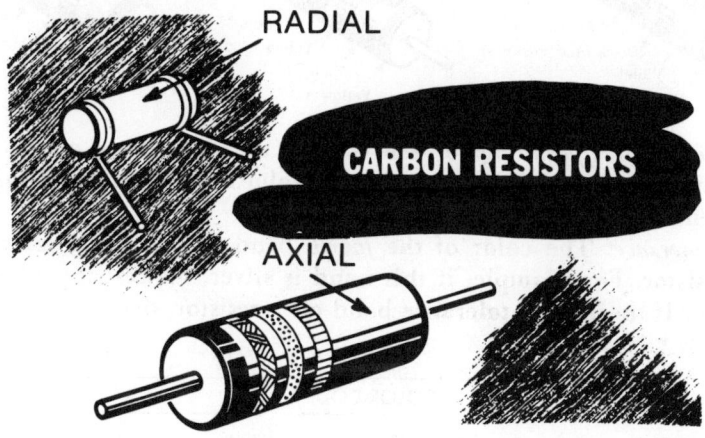
CARBON RESISTORS — RADIAL / AXIAL

Axial-lead resistors are made with the leads molded into the ends of the carbon rod of the resistor body. The leads extend straight out from the ends and in line with the body of the resistor. The carbon rod is completely coated with a material which is a good insulator.

In the color code system of marking, three colors are used to indicate the resistance value in ohms, and a fourth color is sometimes used to indicate the tolerance of the resistor. By reading the colors in the correct order and by substituting numbers from the color code, you can immediately tell all you need to know about a resistor. As you practice using the color code shown on the next page, you will soon get to know the numerical value of each color, and you will be able to tell the value of a resistor at a glance.

Resistor Color Code (continued)

First Significant Figure: On the resistor, the color of the *first* band indicates the *first* digit of the resistance value. For example, as shown in the Color Code Table below, if this band is brown, the first digit is 1.

Multiplying Value: The color of the *third* band indicates the value by which the first two digits are to be multiplied to obtain the resistance value. For example, again using the Color Code Table, if this band is yellow, the first two digits are multiplied by 10,000. (Therefore, with first and second significant digits of 15, the value is 150,000.) This band can also be thought of as indicating the number of zeros to be added after the second digit. When used this way, the number of zeros shown in the Significant Figures column of the Color Code Table is the number of zeros to add. For example, if this band is blue, add six (6) zeros after the *second* digit; but if the band is black, no zeros are added. If the third band is gold or silver, the multiplying value must still be used.

Examples of the use of this table are as follows:

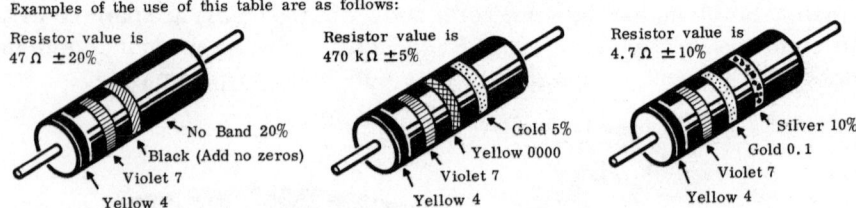

Resistor value is 47 Ω ±20%
No Band 20%
Black (Add no zeros)
Violet 7
Yellow 4

Resistor value is 470 kΩ ±5%
Gold 5%
Yellow 0000
Violet 7
Yellow 4

Resistor value is 4.7 Ω ±10%
Silver 10%
Gold 0.1
Violet 7
Yellow 4

Second Significant Figure: The color of the *second* band on the resistor indicates the *second* digit of resistance value. For example, if this band is green, the second digit is 5.

Tolerance: The color of the *fourth* band indicates the tolerance of the resistor. For example, if this band is silver, the resistor tolerance is ±10%. If there is *no* tolerance band on a resistor, the tolerance is *automatically* ±20%.

COLOR CODE TABLE			
COLOR	Significant Figures	Multiplying Value	Tolerance
Black	0	1	—
Brown	1	10	—
Red	2	100	—
Orange	3	1000	—
Yellow	4	10000	—
Green	5	100000	—
Blue	6	1000000	—
Violet	7	10000000	—
Gray	8	100000000	—
White	9	1000000000	—
Gold	—	0.1	± 5%
Silver	—	0.01	±10%
No Band	—	—	±20%

How Resistance Is Measured

Voltmeters and ammeters are meters you are familiar with and may have used to measure voltage and current. Meters used to measure resistance are called *ohmmeters*. These meters differ from ammeters and voltmeters particularly in that the scale divisions are not *equally* spaced, and the meter requires a built-in battery for proper operation. When using the ohmmeter, no voltage should be present across the resistance being measured except that of the ohmmeter battery; otherwise, the ohmmeter will be damaged.

Ohmmeter ranges usually vary from 0-1,000 ohms to 0-10 megohms. The accuracy of the meter readings decreases at the maximum end of each scale, particularly for the megohm ranges, because the scale divisions become so closely spaced that an accurate reading cannot be obtained. Unlike other meters, the zero end of the ohmmeter scale is at full-scale deflection of the meter pointer.

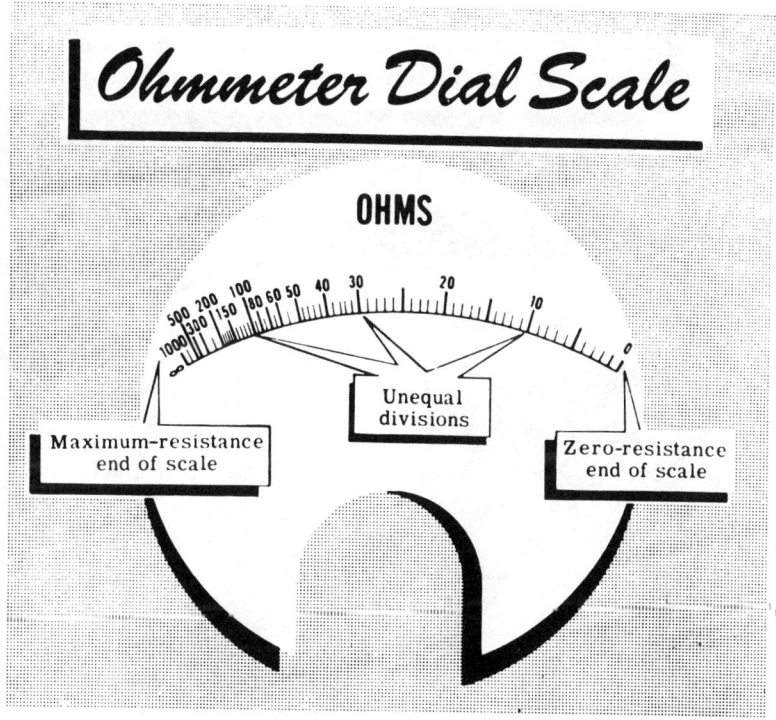

Special ohmmeters called MEGGERS® are required to measure values of resistance over 10 megohms, since the built-in voltage required is very high for ranges above 10 megohms. Some MEGGERS® use high-voltage batteries and others use a special type of hand generator to obtain the necessary voltage. While ohmmeters are used to measure the resistance values of resistors, the most important use of these megohmmeters is to measure and test the resistance of insulation.

How Resistance Is Measured (continued)

You will learn more about the way the ohmmeter works in detail later, when you have learned more about voltage, current flow, and resistance—and their relationship to each other. For now, we will concentrate on how the ohmmeter is used to measure resistance. The principle of the ohmmeter is quite simple. The current through the unknown resistor is measured under conditions where a *known* voltage is applied across the *unknown* resistor. If you have had an opportunity to see an ohmmeter, you may have noticed that in addition to the meter calibrated in ohms, there is a zero adjustment control and a range selector switch. The range selector switch is marked R, R × 10, R × 100, R × 1,000 (or R×1 K), etc. The function of the range selector switch is very much like that of the selector switch on the multirange voltmeters and ammeters that we studied earlier; that is, it allows the selection of the range that gives a reading on the useful part of the scale. It differs, however, in that the position of the range selector switch gives a *multiplying* factor to the values read on the ohmmeter scale. For example, if the range selector switch is on R × 100, then the value read on the meter is multiplied by 100 to get the actual value of the resistance that is being tested.

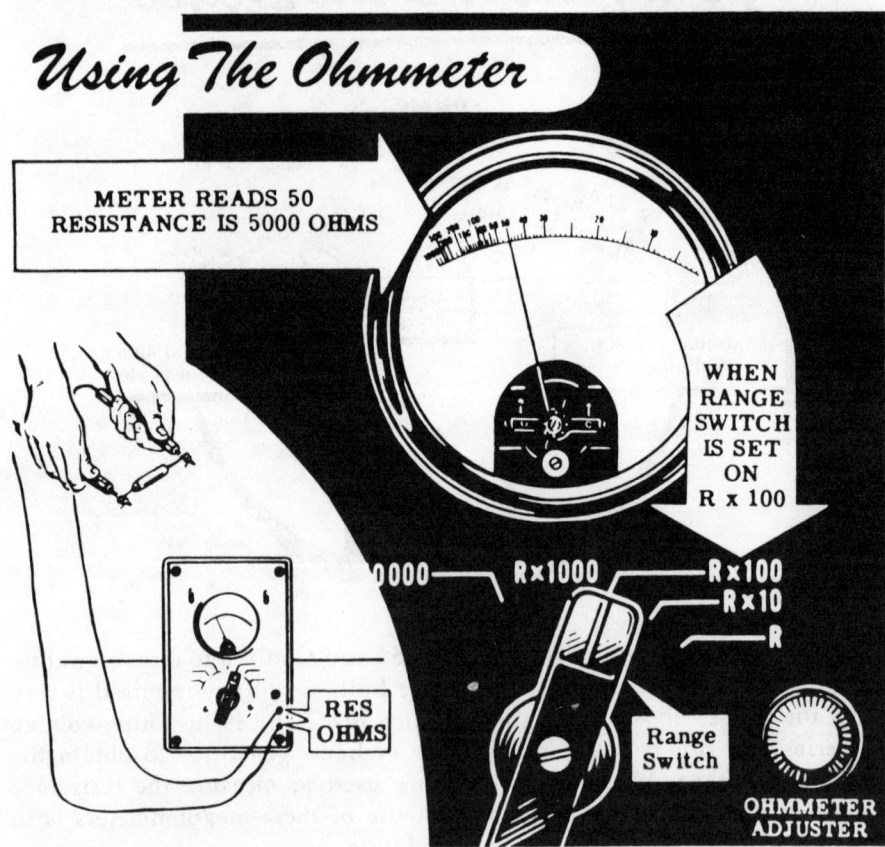

Using The Ohmmeter

METER READS 50
RESISTANCE IS 5000 OHMS

WHEN RANGE SWITCH IS SET ON R x 100

R x 1000
R x 100
R x 10
R

RES OHMS

Range Switch

OHMMETER ADJUSTER

How Resistance Is Measured (continued)

Using the ohmmeter is very simple and proceeds in two steps. First, the voltage must be set to the proper value. This is done with the zero adjustment by shorting out (connecting together) the leads from the ohmmeter and setting to zero ohms on the meter with the zero adjustment. This must be done whenever the meter range selector switch is changed to a different scale. The meter is now calibrated for the given range scale since, with the leads shorted out, the meter reads zero ohms (no resistance between the test leads); and with the test leads open, the meter will indicate infinity (or open circuit). When the unknown resistance is connected between the test leads, the resistance can be read directly from the meter and multiplied by the multiplying factor from the range selector switch.

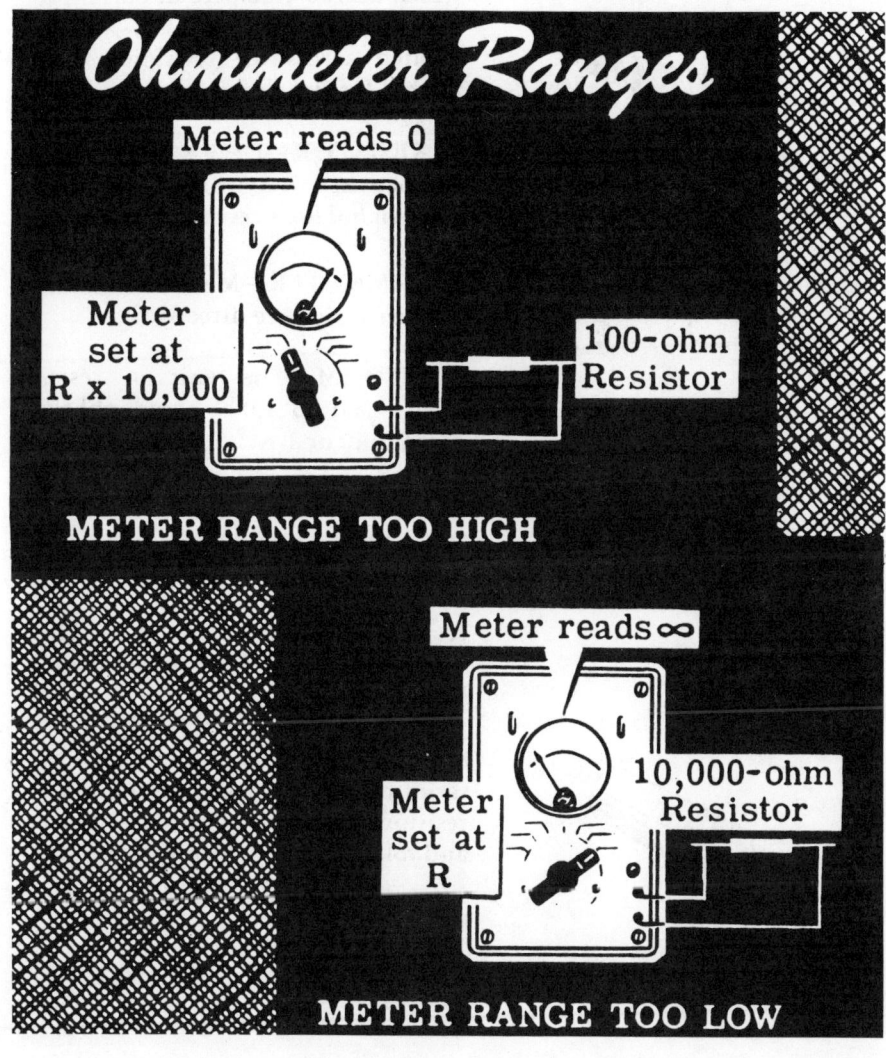

Review of Resistance (Including Material from Volume 1)

You have now learned about the fundamentals of voltage, current, and resistance; and you are ready to go on and see how electric circuits work. Before we go on, let's briefly review what you have learned about resistance and how it is measured.

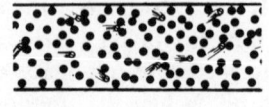

1. RESISTANCE—Opposition offered by a material to the flow of current.

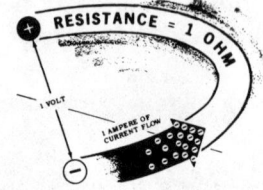

2. OHM—Basic unit of resistance measure equal to that resistance which allows 1 ampere of current to flow when an emf of 1 volt is applied across the resistance. The symbol for the ohm is Ω.

3. RESISTOR—Device having resistance used to control current flow. The symbol for a resistor is "R."

4. OHMMETER—Meter used to measure resistance directly.

1 K = 1,000 ohms

5. KILOHM—The unit of resistance equal to 1,000 ohms, abbreviated 1 kΩ or 1 K.

1 M = 1,000,000 ohms

6. MEGOHM—The unit of resistance equal to 1,000,000 ohms, abbreviated 1 MΩ or 1 M or 1 meg.

7. RESISTOR TOLERANCE—The spread in value of a resistor about its designated value.

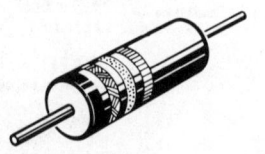

8. RESISTOR COLOR CODE—A series of colored bands on a carbon resistor that tells what the resistance and tolerance of the resistor are.

RESISTANCE

Self-Test—Review Questions (Including Material from Volume 1)

1. Define what resistance is. What is a resistor? What is the symbol used to designate a *fixed* resistor? A *variable* resistor?
2. In a circuit with *constant voltage*, what happens to the current when the resistance is doubled? Halved? Tripled?
3. In a circuit with *constant resistance*, what happens to the current when the voltage is doubled? Halved? Quadrupled? Tripled?
4. Define the unit of resistance. What symbol is used to designate it?
5. What factors determine the resistance of a resistor? Give examples of their effect.
6. Calculate the following conversions using appropriate symbols where applicable.

Convert to Ohms	Convert to Kilohms	Convert to Megohms
6.2 kilohms	4,700 ohms	1,000 kilohms
6.2 megohms	8.2 megohms	120,000 ohms
270 milliohms	100,000 ohms	92,000 ohms
3.3 kilohms	0.1 megohm	68 kilohms
9.1 kilohms	0.39 megohm	470,000 ohms
4.7 megohms	24,000 ohms	330 kilohms

7. What are the preferred values of resistance in 20% tolerance?
8. What is the device used to measure resistance? Describe very briefly how it is used.
9. What are the values of resistance indicated by the following color codes?

Band 1	Band 2	Band 3	Band 4	Resistance Value
red	red	red	gold	
white	brown	yellow	gold	
brown	black	orange	none	
brown	black	black	silver	
violet	green	blue	gold	
grey	red	brown	silver	

10. Describe the steps involved in calibrating and using an ohmmeter.

Learning Objectives—Next Section

Overview—You are now prepared to start using Ohm's Law seriously to solve for current, voltage and/or resistance. In the next section you will start by learning how to solve the simple series circuit.

The Series Circuit

A *series* circuit is formed when two or more resistors are connected *end-to-end* in a circuit in such a way that there is *only one path* for current to flow.

You already know how to connect cells in series so as to form a battery. Connecting resistors in series so as to form a series circuit is even easier. Resistors (unlike cells) have *no* polarity, so you do not have to worry about not connecting two positive or two negative terminals to one another.

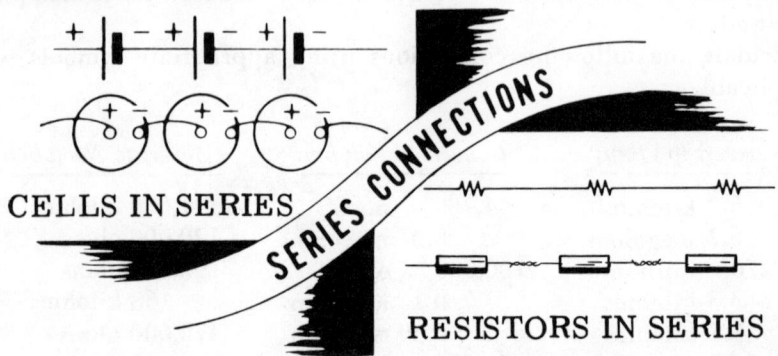

CELLS IN SERIES

SERIES CONNECTIONS

RESISTORS IN SERIES

Note that if you connect a terminal of one lamp socket to a terminal on another socket, leaving one terminal on each socket unconnected, lamps placed in these sockets would be *series-connected*—but you would not have a series circuit. To complete the series circuit, you would have to connect the lamps *across a voltage source*, such as a battery, using the unconnected terminals to complete the circuit.

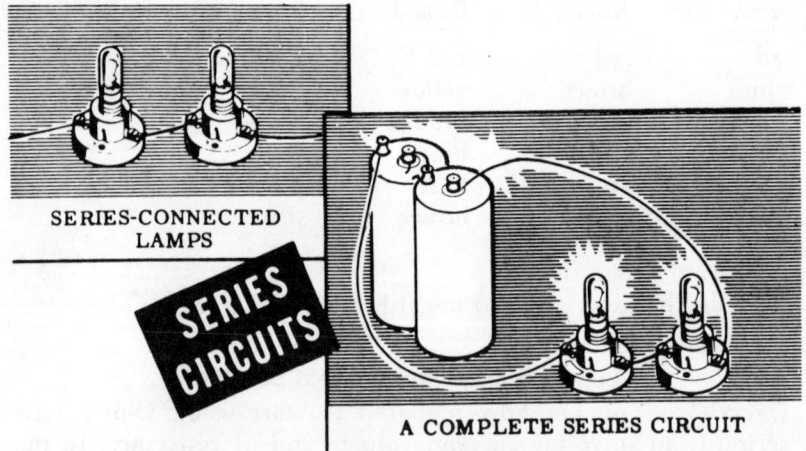

SERIES-CONNECTED LAMPS

SERIES CIRCUITS

A COMPLETE SERIES CIRCUIT

Any number of lamps, resistors, or other devices having resistance can be used to form a series circuit, provided they are connected end-to-end across the terminals of a voltage source and offer only one path for current flow between these terminals.

DC SERIES CIRCUITS

Resistance in Series Circuits

The important thing to remember about resistances connected in series is that *their values add*.

You already know that the resistance of a conductor increases as the length of the conductor increases. It is easy to see, then, that if you connect one length of wire to another, the resistance of the full length of wire will be equal to the sum of the resistances of the original lengths.

For example, if two lengths of wire—one having a resistance of 4 ohms and the other a resistance of 5 ohms—are connected together, the total resistance between the unconnected ends is 9 ohms. Similarly, when other types of resistances are connected in series, the total resistance always equals the sum of the individual resistances.

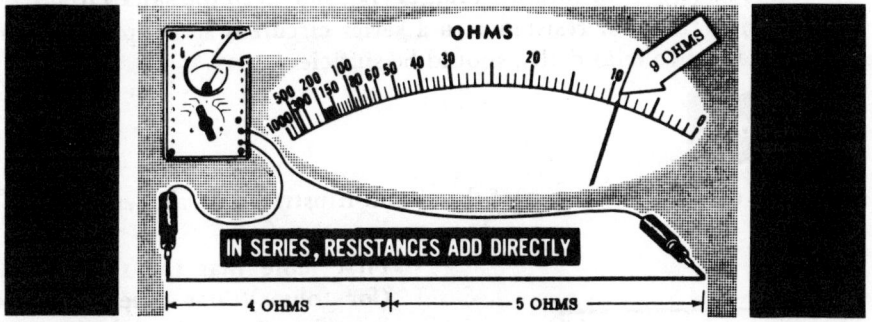

Whenever you use more than one of the same device or quantity in an electric circuit, some method of identifying each individual device or quantity is necessary. For example, if three resistors of different values are used in a series circuit, something other than just R is needed to distinguish them from each other.

To meet this need, a system of numerical identification is used. It consists of following the symbol of the device or quantity by an identification number (or reference). These numbers are sometimes called *subscripts*, because you will occasionally find them written somewhat smaller and slightly offset (below the line), particularly in older drawings. In most modern circuit diagrams, these numbers are written on the line; thus R1 is the same as R_1. R1, R2, and R3 are all symbols for resistors, but each identifies only one *particular* resistor. Similarly, E1, E2, and E3 are all different reference designations for values of voltage used in the same circuit, with the number identifying the particular voltage referred to.

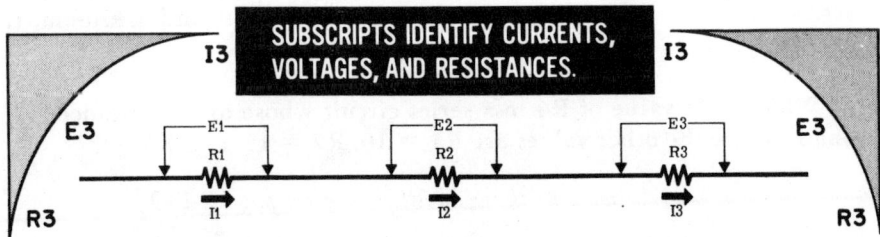

Resistance in Series Circuits (continued)

A small subscript letter *t* is often used to indicate the *total* resistance (R_t) of a series circuit in which two or more individual resistors are included. You know that the total resistance of a series circuit is equal to the sum of the individual resistances in the circuit. In other words,

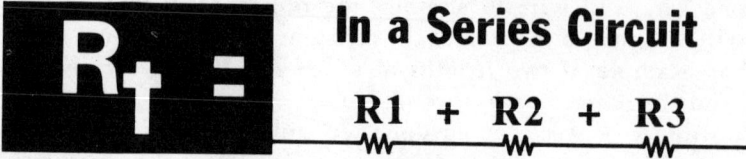

$$R_t = R1 + R2 + R3$$ In a Series Circuit

You will find that the subscript t is also used to express the total of several currents (I_t), or of several voltages (E_t), shown in the same circuit.

Finding the total resistance in a series circuit is so simple that one example, and two trial drills, should be sufficient.

Example

Find the total resistance of the circuit illustrated on the left, below.

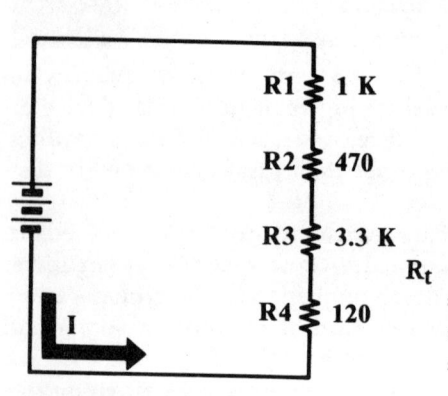

First, note that the symbol "Ω" (for ohms) has been omitted opposite the 470 and the 120. This is often done in circuit diagrams to save space, when the meaning of the figures by themselves is obvious.

Now write your formula:

R_t = R1 + R2 + R3 + R4
  = 1K + 470 + 3.3K + 120
  = 1,000 + 470 + 3,300 + 120 ohms
  = 4,890 ohms

Drill

1. What is the total resistance of a circuit in which three resistors are series-connected with values of 220 ohms, 680 ohms, and 1 kilohm, respectively?

2. What is the value of R4 in a series circuit whose total resistance is 67 ohms, where the other values are R1 = 10, R2 = 15, and R3 = 27?

(Answers to these questions are on page 2-149.)

Current Flow in Series Circuits

You already know that in a series circuit there is only *one* path for current flow, and this means that *all* the current must flow through *every* component in the circuit. If you connect an ammeter onto either end of every resistor in a series circuit, it will show that *an identical amount of current is flowing through every component in the circuit.*

This will not surprise you if you remember how an electric cirrent resembles the electron streaming orbit revolving through the completed circuit, but it is the central fact to grasp before you start applying the principles of Ohm's Law to a series circuit.

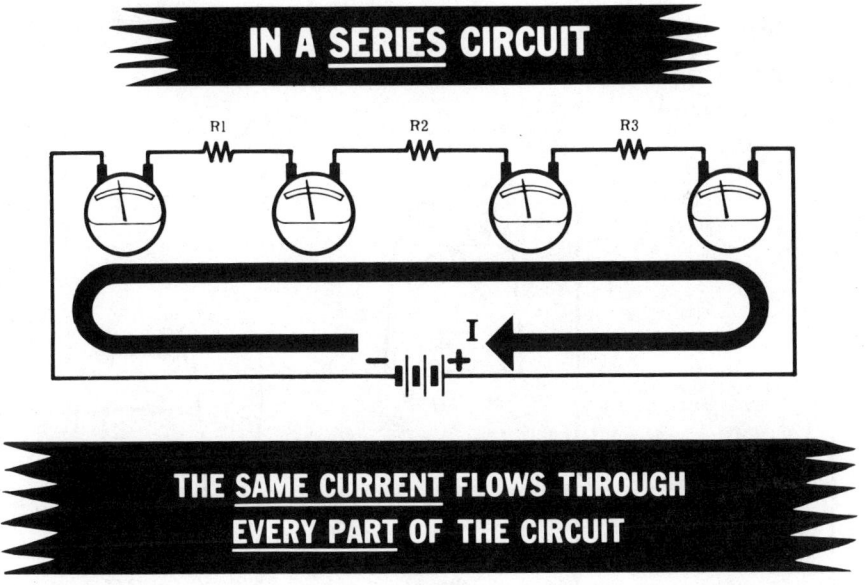

Note, in passing, an important practical consequence of the principle illustrated above. The fact that *all* the circuit current must pass through *every* part of the series circuit means that *every component* connected into such a circuit must be capable of passing, without being damaged, the current which will flow through the circuit.

Lamps connected in series, for example, must all be rated to pass the full circuit current. When rated too low, they will light very brightly, and they will also be liable to burn out because of the excessive current flowing through them.

It is important to remember that *exactly the same thing can happen* if the circuit, instead of lamps, contains resistors, whose shorting out could be even more inconvenient or costly. A resistor required to pass more than its rated current will get extremely hot. Eventually, it will fail completely and become an open circuit. The equipment of which it forms a part will then probably cease to function, and trouble is at hand!

Voltages in Series Circuits—Kirchhoff's Second Law

You know that when a voltage moves electrons through a resistance, some of the available emf is used up. Such a loss of emf is called a *potential drop* or a *voltage drop* across that resistance. You will now find out how this voltage drop is distributed between several resistors of equal value connected in series.

Connect three resistors of equal value in series across a 6-volt battery, and touch voltmeters across the points as shown in the diagram below. Since the current passing through each of the equal resistors is the same, the energy expended in pushing this equal amount of current through each individual resistor must also be the same.

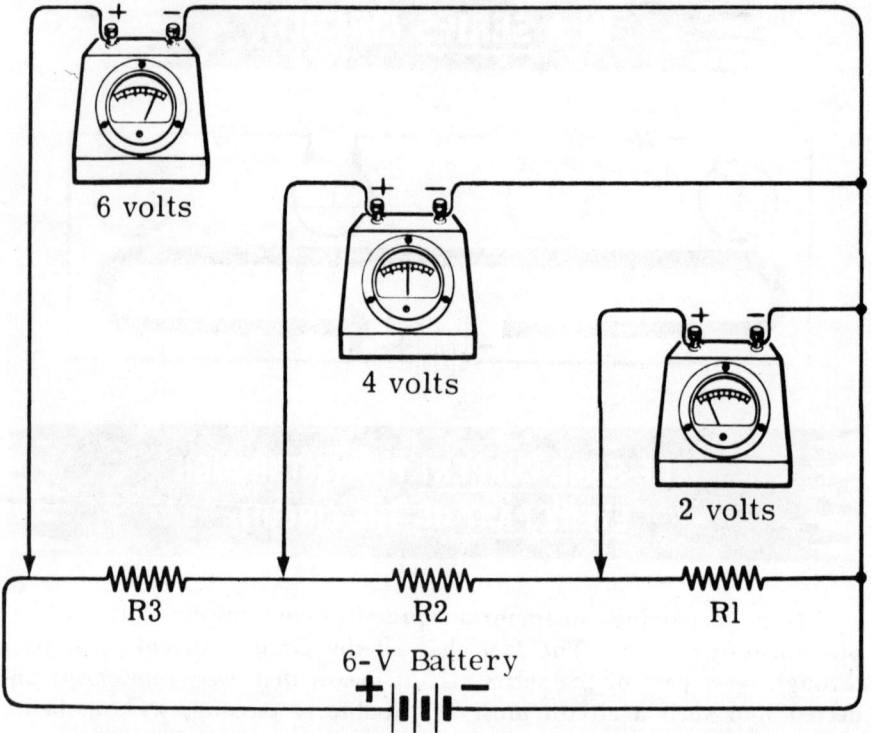

In other words, the voltage across each of the three resistors pictured above is 2 volts. The voltage drop across R1 will therefore read on your voltmeter as 2 volts; that across R1 and R2 in combination will read 4 volts; and that across R1, R2, and R3 (the complete circuit) will read 6 volts. If you add together the voltage drops across all three resistors, you will get exactly the original supply voltage (6 volts).

This important fact was expressed by the German physicist Kirchhoff (1824-1887) in what is known as Kirchhoff's Second Law. (We will be discussing his First Law shortly.)

Kirchhoff's Second Law states: *The sum of the voltage drops across the resistances of a closed circuit equals the total voltage applied to the circuit.*

Ohm's Law in Series Circuits

You now know three important facts about a series circuit:

1. The current flowing through it is the same everywhere. This can be expressed by the equation $I_t = I_1 = I_2 = I_3$, and so on.

2. The total resistance of the circuit equals the sum of the individual resistances in it. This can be expressed by the equation $R_t = R_1 + R_2 + R_3$, and so on.

3. When the voltage drops in a series circuit are added together, their total value is equal to the total applied voltage (Kirchhoff.) This can be expressed by the equation $E_t = E_1 + E_2 + E_3$, and so on.

These three facts, used in conjunction with Ohm's Law, will be of constant help to you in determining the values of complete circuits, or parts of circuits, on the frequent occasions when you will either lack the correct meter to tell you the answer directly, or find it impossible to use a meter to obtain a value directly when constructing a circuit.

This will happen to you time and time again when you get on to the really interesting applications of electrical and electronic principles.

One useful simplification arises from the equation, $R_t = R_1 + R_2 + R_3$, etc. Look at the two circuits below, and you will quickly see that the right-hand equation is a more convenient *equivalent* version of the one on the left.

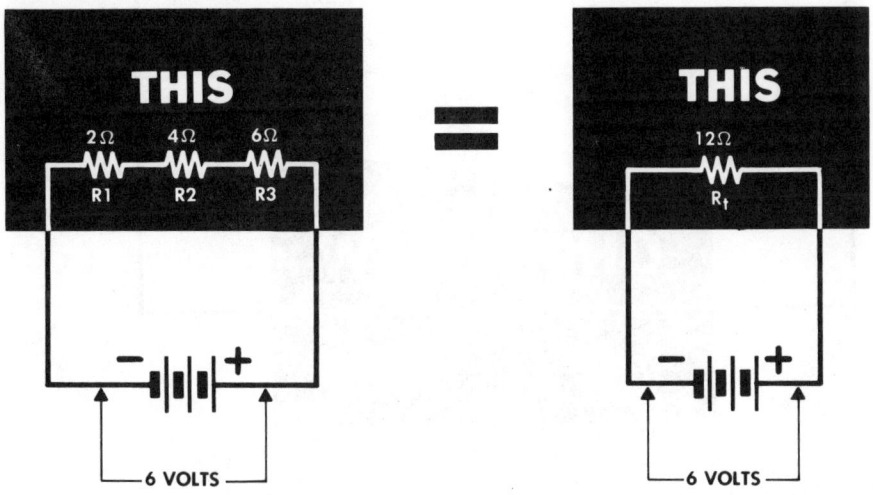

You can soon find the missing factor—the current through either circuit—by using Ohm's Law and the known facts that E=6 and R=12. The equation you want is (thumb on the I of the magic triangle!)

$$I = \frac{E}{R} = \frac{6}{12} = 0.5 \text{ ampere}$$

Always be on the lookout for a chance to simplify the resistances in a series circuit into the single resistance of an *equivalent circuit*.

Ohm's Law in Series Circuits (continued)

Ohm's Law can be usefully applied in series circuits either to the complete circuit itself or to individual parts of the circuit. Together with Kirchhoff's Second Law, for instance, it will enable you, by calculation, to insert a great many missing values in a series circuit of the following kind:

A circuit contains three resistors connected in series across 100 volts; the circuit current flow is 2 amperes. Two of the resistors (call them R1 and R2) have known values of 5 and 10 ohms, respectively. You wish to know the resistance of the entire circuit, the value of the third resistor, R3, and the voltage drops across each of the three resistors.

First, sketch the circuit on paper. Fill in the values you already know, leaving blanks opposite the ones you don't know.

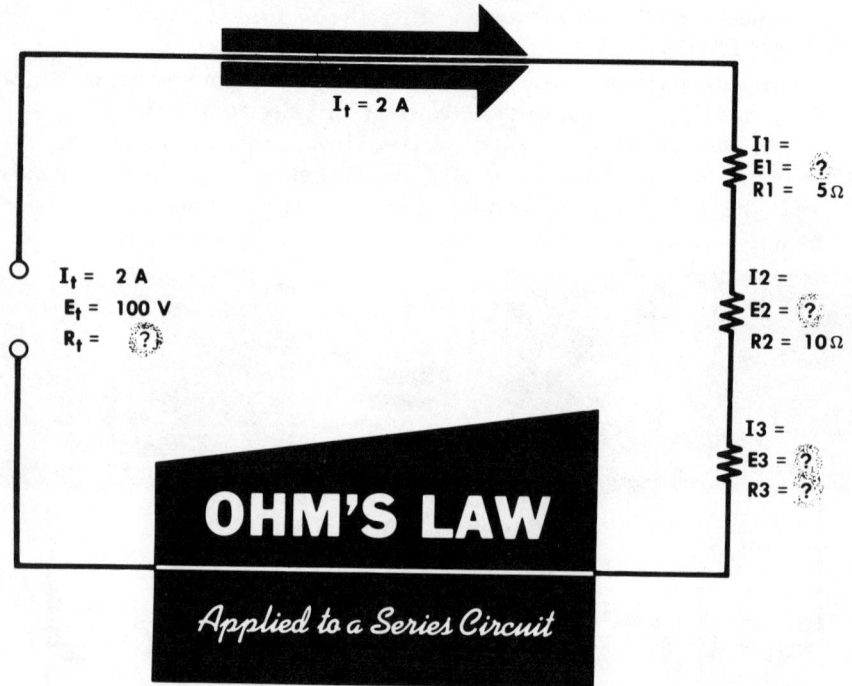

Now, for the missing values. The first and most obvious one is the current. You know that $I_t = I_1 = I_2 = I_3$. . . . You are quite safe in filling in the value of 2 amperes for I wherever it appears on your sketch.

Next, take the group of figures on the left of your sketch. Finding the value of R_t is now a simple exercise with Ohm's Law. With your thumb *mentally* on R in the magic triangle, you get

$$R_t = \frac{E_t}{I_t} = \frac{100}{2} = 50 \text{ ohms}$$

Ohm's Law in Series Circuits (continued)

Now that you know the total circuit resistance is 50 ohms, you can easily fill in the missing link in the equation:

$$R_t \ (50\Omega) = R1 \ (5\Omega) + R2 \ (10\Omega) + R3 \ (?)$$

Rewrite the equation as $R3 = R_t - R1 - R2$, and you get

$$R3 = 50 - 5 - 10 = 35 \text{ ohms}$$

Fill in this value on your sketch (see previous page) and you are clearly making progress.

Now, consider what you know about the resistor, R1. You know its value to be 5 ohms, and that the current through it is 2 amperes. Ohm's Law will give you the voltage drop across it. So *Thumb on E!* and

$$E1 = I_1 \times R1 = 2 \times 5 = 10 \text{ volts}$$

Do the same thing for E2, and you get

$$E2 = I_2 \times R2 = 2 \times 10 = 20 \text{ volts}$$

Fill both these values in on your sketch, and the only gap left in the entire list of circuit values is E3.

This last value can be calculated in two different ways. Use both methods, for they not only provide a useful check on one another, but will also prove that Ohm's Law and Kirchhoff's Second Law work accurately!

First, Ohm's Law tells you that

$$E3 = I_3 \times R3 = 2 \times 35 = 70 \text{ volts}$$

Then, Kirchhoff's Second Law tells you that $E_t = E1 + E2 + E3$. Transposing, you get

$$E3 = E_t - E1 - E2 = 100 - 10 - 20 = 70 \text{ volts}$$

Ohm's Law in Series Circuits (continued)

Practice makes perfect—and it is so important that you grasp the use of Ohm's Law in solving series circuits that you should work through the exercise which follows (it is very similar to the one you did on the last two pages) and then tackle a page of problems like it on your own.

Example

In a circuit of which the relevant part is shown below, you measure a voltage drop of 5 volts across R1; but you cannot get your voltmeter leads across R2 and R3.

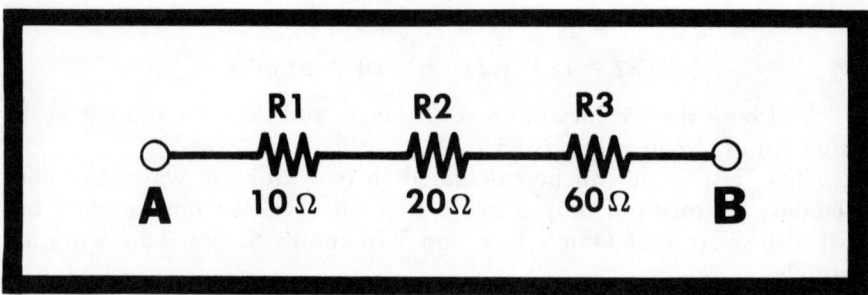

Calculate the voltage drops across R2 and R3. What is the total voltage applied across Points A-B?

You know two facts about R1—its value, and the voltage drop across it. So you can use Ohm's Law to find the current through it.

$$I = \frac{E}{R} = \frac{5}{10} = \frac{1}{2} \text{ ampere}$$

Given the current anywhere in a series circuit, you automatically know it everywhere throughout the circuit; so you can readily calculate E2 as

$$E2 = I_2 \times R2 = 1/2 \times 20 = 10 \text{ volts}$$
$$E3 = I_3 \times R3 = 1/2 \times 60 = 30 \text{ volts}$$

You have now calculated the voltage drops across E2 and E3 to be 10 and 30 volts, respectively, and you already know that the voltage drop across E1 is 5 volts. Kirchhoff's Law tells you that the voltage applied across the circuit resistance is the sum of the voltage drops across the individual resistors; thus,

$$E_t = E1 + E2 + E3 = 5 + 10 + 30 = 45 \text{ volts}$$

Further practice on your own will be provided on page 2-46.

Voltage Division in the Series Circuit

You often need to be able to take current from a given point in a series circuit at a stepped-down voltage—that is to say, at a voltage which is lower than the applied voltage.

A circuit widely used for this is shown in the diagram below. It is called a *voltage divider*.

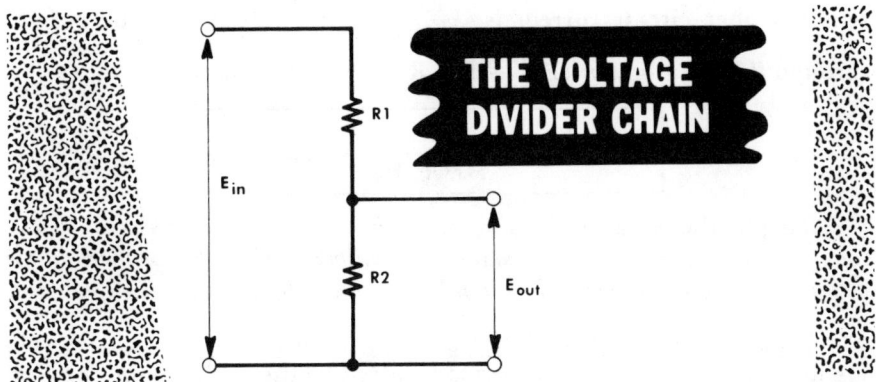

THE VOLTAGE DIVIDER CHAIN

Assume that in this circuit the applied voltage (E_{in}, as it is written in electrical notation) is 100 volts, and that the values of R1 and R2 are 10 and 15 ohms, respectively. What sort of a voltage drop can you expect across R2? (Note that the notation for this voltage is E_{out}. Probably, it will be needed for the *input* voltage into another circuit, in which it will, of course, become E_{in} once more.)

You know that total resistance in this circuit is 10 + 15 = 25 ohms. You also know that the circuit voltage is 100 volts; so you can use Ohm's Law to find the circuit current.

$$I = \frac{E_{in}}{R_t} = \frac{100}{25} = 4 \text{ amperes}$$

Now look at R2. You know that its value is 15 ohms, and you have just calculated that the current through it is 4 amperes (remember that *the current anywhere is the current everywhere* throughout a series circuit). So you get

$$E_{out} = I \times R = 4 \times 15 = 60 \text{ volts}$$

Observe that by the appropriate choice of resistor values in a voltage divider chain, an input voltage of 100 volts has been *stepped down* to an output voltage of 60 volts. And Ohm's Law has enabled you to calculate in advance that this will be so.

Voltage Division in the Series Circuit (continued)

It is obviously possible, from what you have learned on the previous page, to work out a formula for the output voltage of a voltage divider chain which can be applied to a circuit containing a pair of resistors of *any* value.

You can see that total circuit resistance is R1 + R2 and Ohm's Law tells you that circuit current is $\frac{E_{in}}{R1+R2}$. Since this is also the current through R2, Ohm's Law ($E_{out} = I \times R2$) gives the equation:

$$E_{out} = \frac{R2}{R1 + R2} \times E_{in}$$

To put the equation into words: *The voltage across any resistor in a voltage divider chain can be calculated by multiplying the value of that resistor by the input voltage divided by the total resistance of the circuit.*

Note on Thevenin's and Norton's Theorems: It is easy to solve voltage divider problems when a load is involved by use of Thevenin's or Norton's theorem. These theorems, which are discussed on pages 2-133 through 2-136, are useful when you need to obtain a *particular* voltage under *load* conditions.

Variable Resistors

It is frequently convenient to be able to vary the value of a resistor in a circuit at will. (You have often done it yourself when you adjusted the volume control on your radio!)

A common means by which a resistor can be made *variable* in this way is for a sliding arm made of good conducting material to be arranged so it can be moved along the length of the resistor. The resistor is then connected into the circuit with one of its ends fastened to the sliding arm. By moving this sliding arm along the resistor, the value of the resistor can be varied at will between maximum and minimum (zero).

When a variable resistor is used in this way, it is called a *rheostat*. It is generally used to control *current flow* in a circuit.

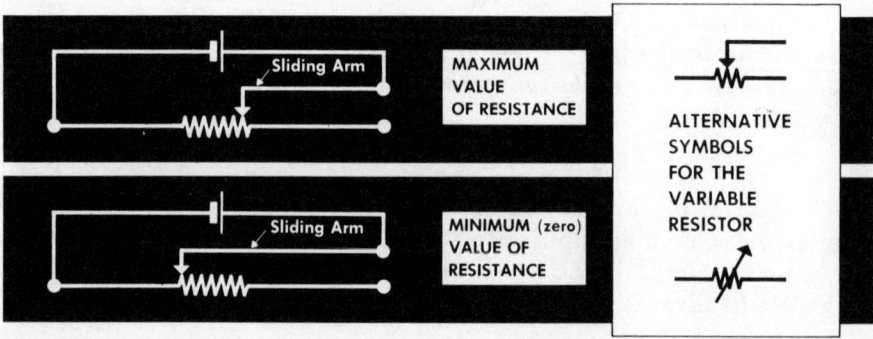

Variable Resistors (continued)

A variable resistor may have either two or three circuit connections. In the diagram below, A and C would always be connected; but B could either be connected or not, at will.

This is what two-terminal and three-terminal resistors, connected as rheostats, look like. The circuit diagram of each is shown below it.

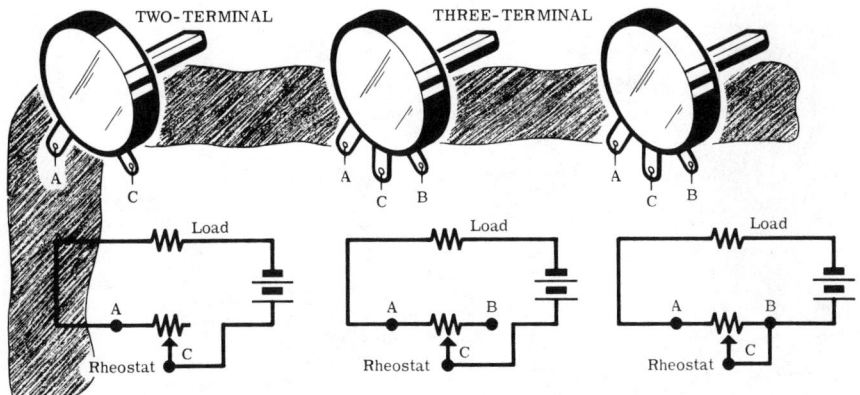

A three-terminal resistor in which all three terminals are connected into the circuit is a *potentiometer*. It is used to control circuit *voltage*.

Potentiometer Connections

The circuit diagram of a potentiometer is really no more than that of a voltage divider chain. R1-R2 is a single resistor, effectively divided by the sliding arm C, whose movement alters the relative values of R1 and R2.

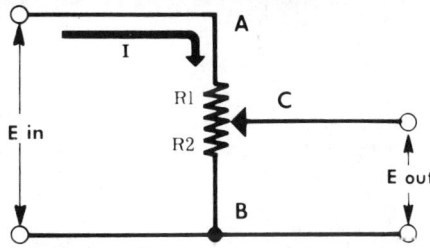

The output voltage can vary from zero (when C is *lowered* so that R2 = 0) to the full circuit voltage (when C is *moved up* so far that R1 = 0).

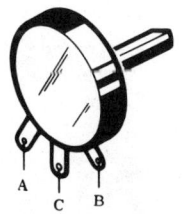

A typical potentiometer looks like this. Note how the connection of all three terminals into the circuit (at points corresponding to A, B, and C in the diagram at the top of the page) enables circuit voltage to be controlled.

Variable Resistors (continued)

Variable resistors, like fixed resistors, can be made with resistance material of carbon or can be wire-wound, depending on the amount of current to be controlled—wire-wound for *large* currents and carbon for *small* currents.

Wire-wound variable resistors are constructed by winding resistance wire on a porcelain or bakelite circular form, with a contact arm which can be adjusted to any position on the circular form by means of a rotating shaft. A lead connected to this movable contact can then be used, with one or both of the end leads, to vary the resistance used.

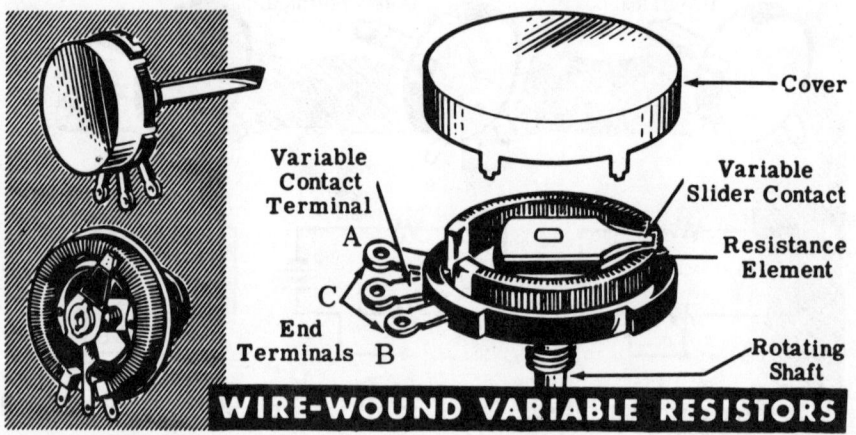

WIRE-WOUND VARIABLE RESISTORS

For controlling small currents, carbon variable resistors are constructed by depositing a carbon compound on a fiber disk. A contact on a movable arm acts to vary the resistance as the arm shaft is rotated.

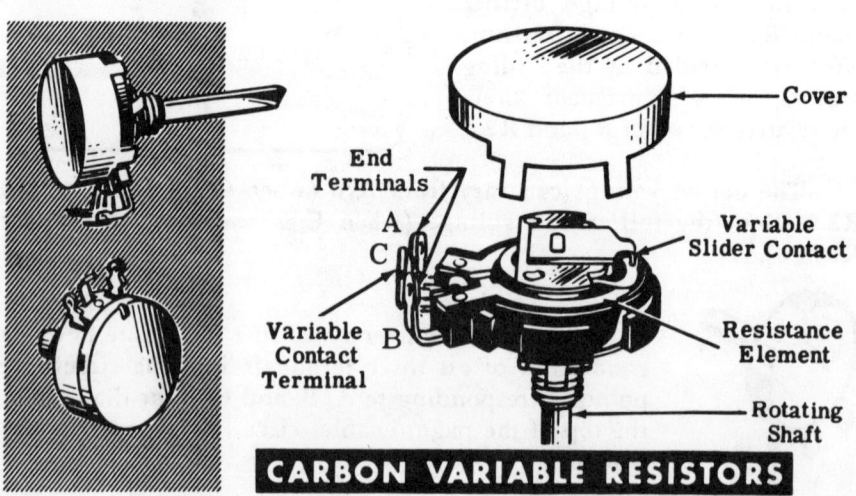

CARBON VARIABLE RESISTORS

Review of Ohm's Law in Series Circuits

1. CURRENT—The value of *current* flowing through a series circuit is always the *same* at every point in the circuit.

$I_t = I_1 = I_2 = I_3 = I_4 = \ldots$

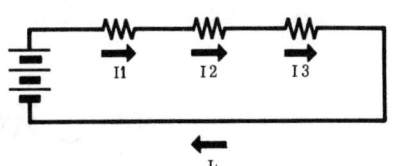

2. RESISTANCE—The total *resistance* in a series circuit is always the sum of the individual values of resistance in the circuit.

$R_t = R1 + R2 + R3 + R4 + \ldots$

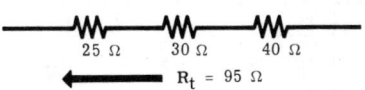

3. VOLTAGE—The *voltage* applied across the circuit resistance of a series circuit is always equal to the *sum* of the voltage drops across the individual resistances.

$E_t = E1 + E2 + E3 + E4 + \ldots$

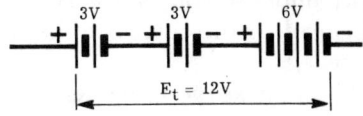

4. UNKNOWNS—The way to find unknown quantities in a series circuit is as follows:
(a) Draw the circuit diagram.
(b) Insert on this diagram all the known facts.
(c) Look for resistances in the circuit about which you know any two values.
(d) Use Ohm's Law to find the third value.
(e) Carry on from your increasing store of known facts to calculate further *unknowns*—never forgetting that once you know the value of any I in the circuit you know the value of all of them.

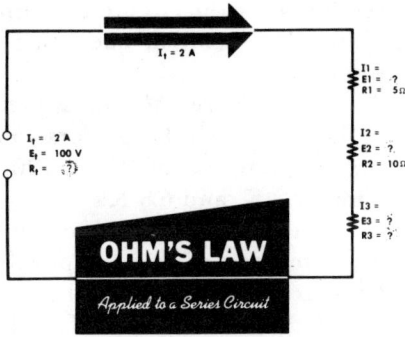

OHM'S LAW
Applied to a Series Circuit

5. EQUIVALENT CIRCUIT—A series circuit containing two or more resistances can often be usefully simplified into an *equivalent circuit* containing a single theoretical resistance having a value equal to the sum of all the actual resistances in the circuit.

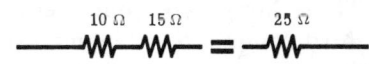

Self-Test—Review Questions

1. What is the total resistance of these four resistors connected in series?

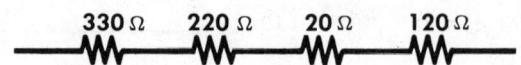

2. Draw the equivalent of the series resistance found in problem 1.
3. (a) Draw a set of 1.5-volt cells in a series circuit so that the total battery voltage is 9 volts. How many cells does it take? Why? Show polarity of connection.
   (b) Suppose you connected one of the cells in part (a) so that it was backwards. What would the total voltage across the battery be? Why?
4. (a) Suppose you had a 100-volt source, but you wanted to get 30 volts instead. If the current to be drawn is 1 A, calculate the divider resistors needed.
   (b) Suppose you wanted a 0-30 volt source. Show how you would change the circuit in part (a) to do this. What would be the resistance of the variable resistor?
5. Draw a series circuit showing two lamps and a battery of 12 volts. Calculate the currents in each lamp and the voltage across each lamp if the resistances are 40 and 60 ohms, respectively.
6. What would happen in question 5 if two 12-volt batteries were put in series? Draw the circuit. Explain.
7. Using the circuit in question 5, show how Kirchhoff's Second Law applies.
8. A 47-kilohm resistor in a piece of equipment fails and you have to replace it. The only resistors you can find are an assorted lot, having values, respectively, of 22 K, 120 K, 15 K, 220 ohms, 4.7 K, 4.7 K, 330 ohms, 1.2 K, and 6.8 K.
   (a) Which of these resistors would you choose to connect in series in order to come as close as possible to the value of 47 K, and what would be their combined resistance?
   (b) If the resistor which failed had a 10% tolerance, would the combination you have chosen be adequate to do its job?
9. Although your car has a 12-volt battery, an aunt gives you a handsome spotlight which takes 2 amperes of current, but which is, unfortunately, designed to work off a 6-volt battery. What values of resistance would you need to connect in series with the spotlight before you could safely switch it on? Draw the circuit diagram and show voltages and current.
10. Three lamps are connected in series. The resistances of the first and third lamps are 52 ohms each; the middle lamp is rated at 76 ohms. What current flows through the lamps when they are connected across a 120-volt supply?

DC SERIES CIRCUITS

Experiment/Application—Open Circuits

You already know that in order for a current to pass through a circuit, a *closed* path (complete loop) is required. Any break in the closed path causes an *open* circuit and stops current flow. Each time you open a switch, you are causing an open circuit.

Anything which causes an open circuit, other than actually opening a switch, interferes with the proper operation of the circuit, and must be corrected. An open circuit may be caused by a loose connection, a burned-out resistor or lamp filament, poor solder joints or loose contacts, or a broken wire.

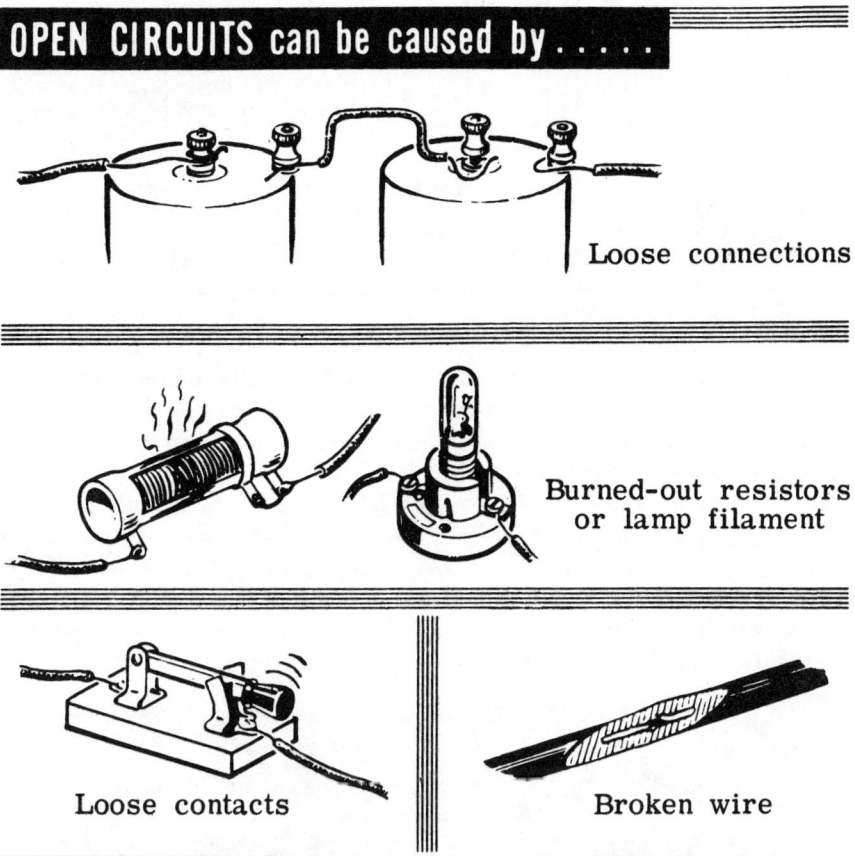

OPEN CIRCUITS can be caused by . . . . .

Loose connections

Burned-out resistors or lamp filament

Loose contacts

Broken wire

These *faults* or *troubles* can often be detected visually, and you may find that in your work you might encounter one or more of these *opens*.

In some cases it is not possible to detect visually the cause of an open circuit. An ohmmeter or a test lamp can then be used to find the cause of the trouble.

Experiment/Application—Open Circuits (continued)

Now, suppose you connect five dry cells, a knife switch, and three lamp sockets in series. Insert three 2.5-volt, 0.75-ampere lamps in the sockets. When you close the switch, the lamps light with normal brilliancy. If you then loosen one of the lamps, they all go out, indicating an *open* circuit. (A loosened lamp simulates a burned-out filament or other open.)

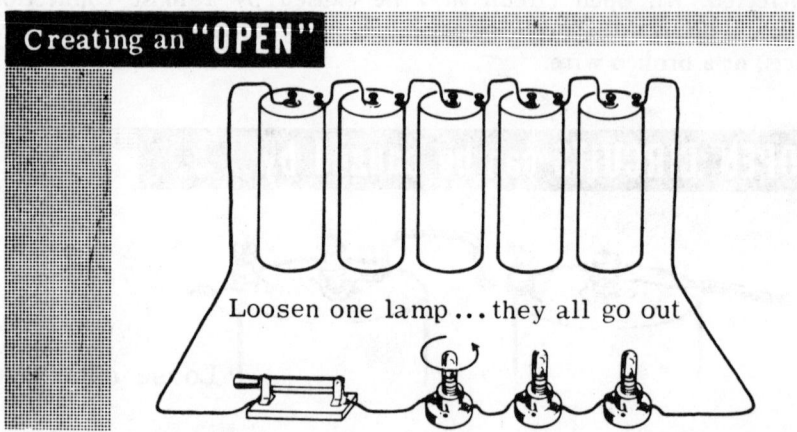

Creating an "OPEN"

Loosen one lamp... they all go out

To locate the open with the ohmmeter, you would first open the knife switch to remove the voltage source, since an ohmmeter must *never* be used on a circuit with the power connected. Then touch the ohmmeter test leads across each unit in the circuit—the three lamps in this case. You would see that for two of the lamps, the ohmmeter indicates a resistance of about 4 ohms; but for the loosened lamp, the ohmmeter indicates *infinity*. Since an open does not allow any current to flow, *its resistance must be infinite*. The way to use an ohmmeter check for an open, then, is to find the series-connected element in the circuit which measures infinite resistance on the ohmmeter. Remember, on an ohmmeter *zero ohms* is at the *right side* of the scale and *infinity* is on the *left side*.

Using the ohmmeter to test for an "OPEN"

Good bulb—resistance about 4 Ω   "Open"— infinite resistance

Experiment/Application—Open Circuits (continued)

The second method used to locate an open is to test the circuit by means of a test lamp. A test lamp can be set up by attaching leads to the terminals of a lamp socket and inserting a 2.5-volt lamp. If you then close the circuit switch and touch the test lamp leads across each lamp in the circuit, the test lamp will not light until it is across the terminals of the loosened lamp. The test lamp then lights, indicating to you that you have found the open.

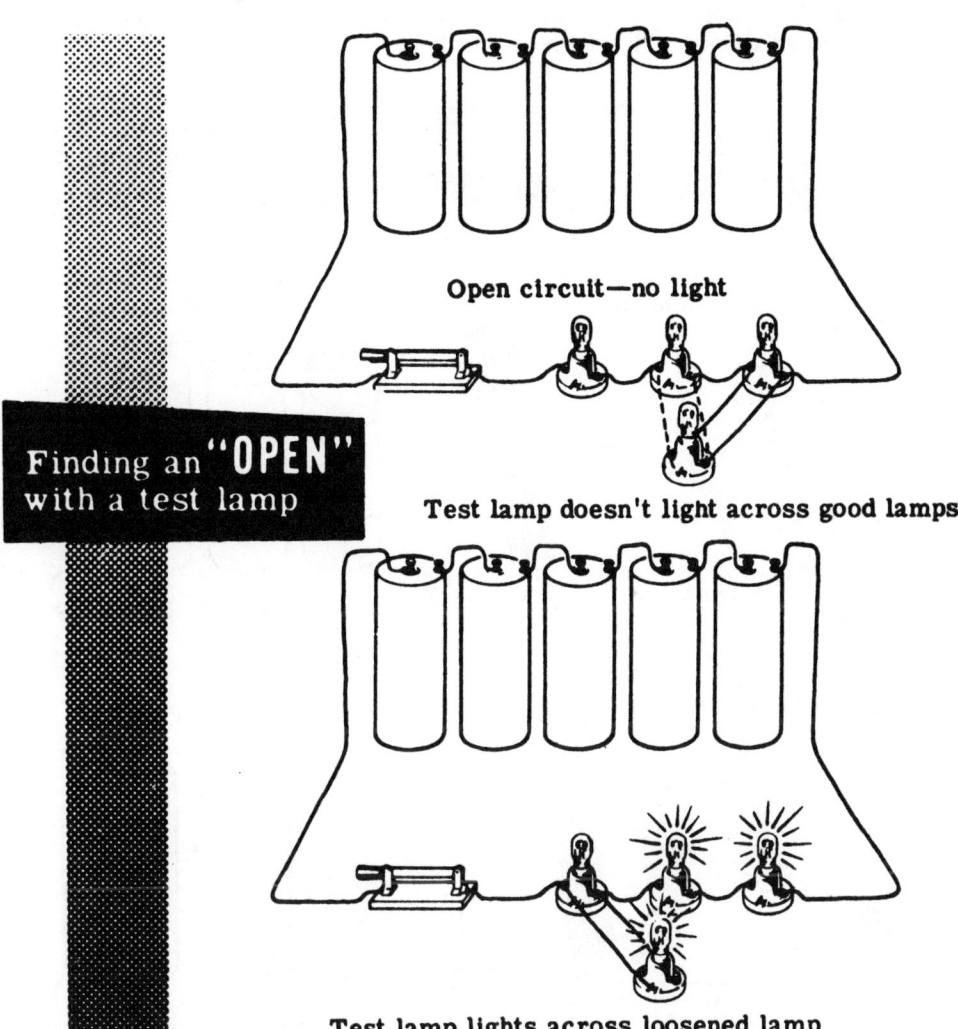

Finding an **"OPEN"** with a test lamp

Open circuit—no light

Test lamp doesn't light across good lamps

Test lamp lights across loosened lamp

The test lamp completes the circuit and allows current to flow, bypassing the open. (This also causes the other lamps to light since the open is being bypassed.) You will often use this method to detect opens which cannot be seen.

Experiment/Application—Short Circuits

You have seen how an open prevents current flow by breaking the closed path between terminals of the voltage source. Now you will see how a *short* produces just the *opposite* effect—creating a *short circuit* path of low resistance through which a larger than normal current flows.

A short occurs whenever the resistance of a circuit or part of a circuit drops from its normal value to a much lower or zero resistance. This happens if the two terminals of a resistance in a circuit are directly connected, the voltage source leads contact each other, two current-carrying uninsulated wires touch, or the circuit is improperly wired.

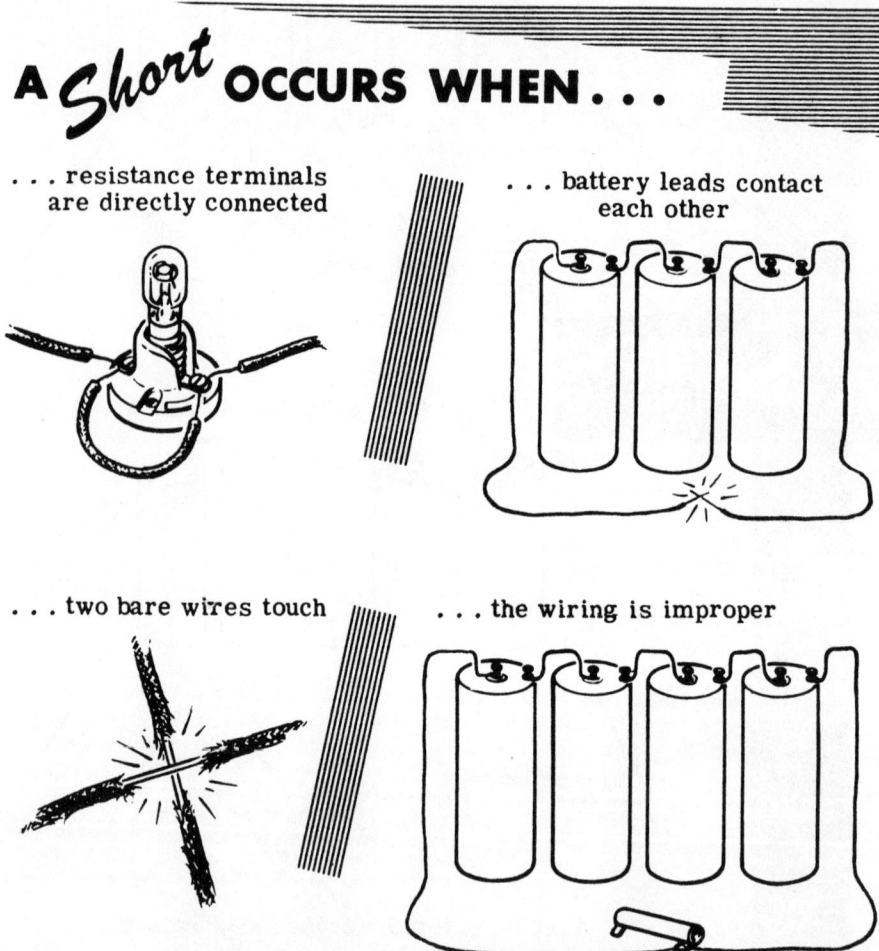

A *Short* OCCURS WHEN...

...resistance terminals are directly connected

...battery leads contact each other

...two bare wires touch

...the wiring is improper

These shorts are called *external shorts* and can usually be detected by visual inspection.

Experiment/Application—Short Circuits (continued)

When a short occurs in a simple circuit, the resistance of the circuit to current flow becomes very low, so that a very large current flows.

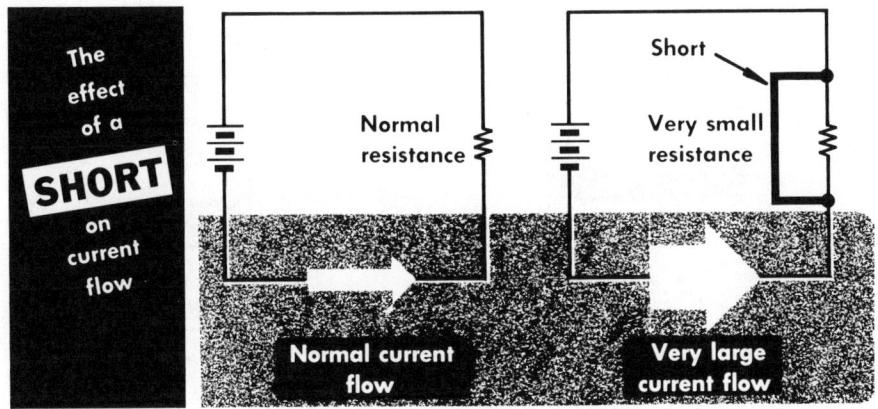

In a series circuit, a short across one or more parts of the circuit results in reduction of the total resistance of the circuit and correspondingly increased current flows, which may damage the other components in the circuit.

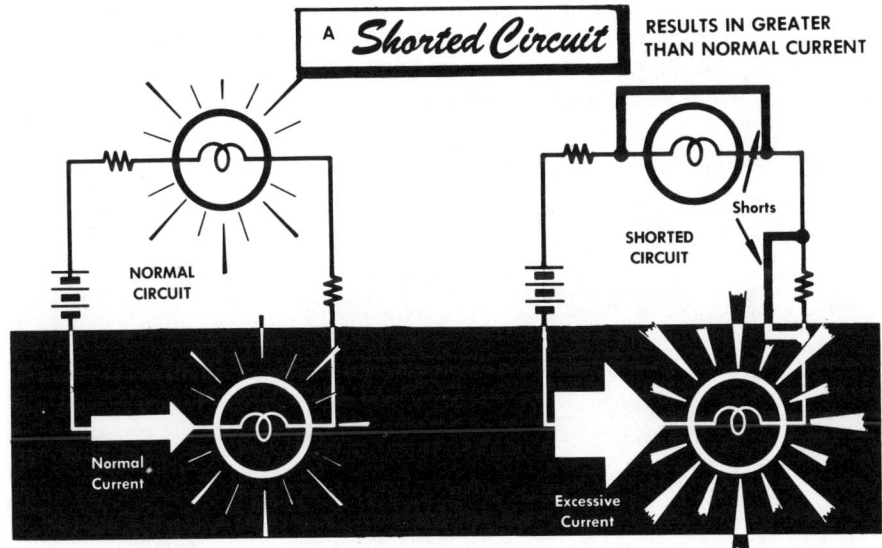

Circuits are usually protected against excessive current flow by the use of fuses, which you will learn about later. But it is important that you understand the reasons for and results of shorts, so that you can *avoid* accidentally shorting your circuits and causing damage to meters or other equipment or components.

Experiment/Application—Short Circuits (continued)

Again, suppose you connect three dry cells in series with a 0-1 ampere range ammeter and three lamp sockets. Then you insert three 2.5-volt, 0.75-ampere lamps in the sockets and close the switch. You would see that the lamps light equally but are dim because the voltage is only 6 volts and the ammeter indicates a current flow of about 0.5 ampere.

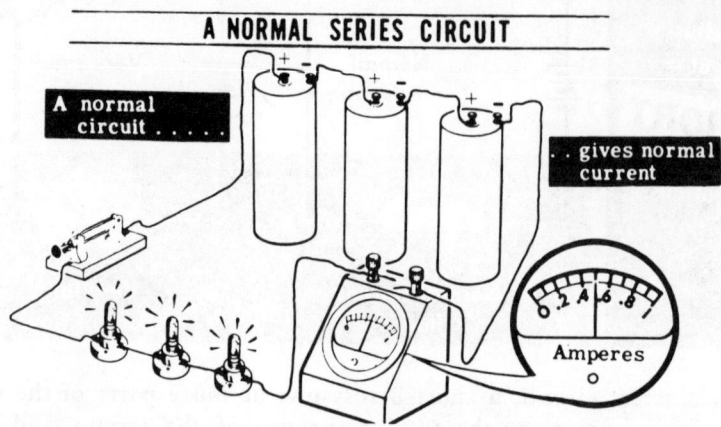

Now suppose you touch the ends of an insulated lead to the terminals of one of the lamps, *short-circuiting* the current around that lamp. You would see that the lamp goes out, the other two lamps become brighter, and the ammeter shows that the current has increased to about 0.6 ampere. If you move the lead to short out two of the lamps, you would see that they both go out, the third lamp becomes very bright, and the current increases to about 0.9 ampere. Since the lamp is rated at only 0.75 ampere, this excessive current would soon burn out the filament.

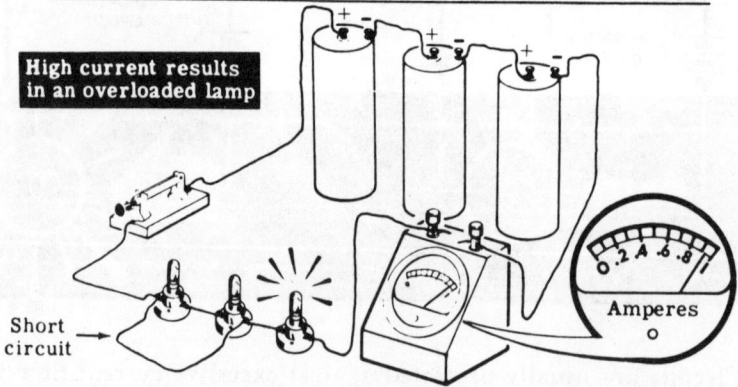

If you were to short out all three lamps, the lack of resistance of the circuit would cause a great amount of current to flow, which would *damage* the ammeter.

Experiment/Application—Series Circuit Resistance

You can see the effect of connecting resistances in series by measuring the resistance of three lamps individually, and then measuring the total resistance when they are connected in series.

Suppose you connected three lamp sockets in series and inserted a 6-volt, 0.5-ampere lamp in each socket. By using an ohmmeter to measure the resistance of each lamp, you would see that each lamp resistance measures about 12 ohms.

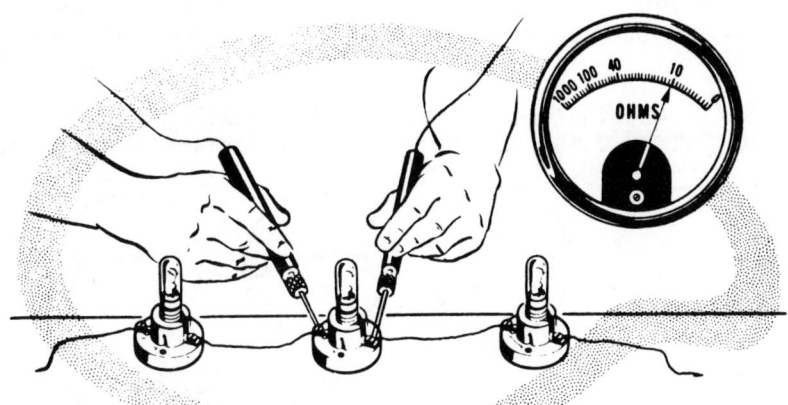

Next, if you measured the resistance of the three lamps in series, you would see that the total resistance is about 36 ohms. Thus, the total resistance of series-connected resistance is equal to the sum of all the individual resistances.

$$R_t = R1 + R2 + R3$$

Experiment/Application—Series Circuit Resistance (continued)

Now, suppose you connect four dry cells in series to form a 6-volt battery as a voltage source. Connect the battery, one lamp socket, a 0-1 ampere ammeter, and a switch in series; then connect a 0-10 volt voltmeter across the battery (see illustration below). You would notice that the voltmeter reads 6 volts. If you inserted a 6-volt, 0.5-ampere lamp in the socket and closed the switch, the ammeter would record a current flow of about 0.5 ampere and the lamp would light to normal brilliance.

If you now connected the voltmeter directly across the lamp, instead of across the battery, you would see the voltage across the lamp is 6 volts.

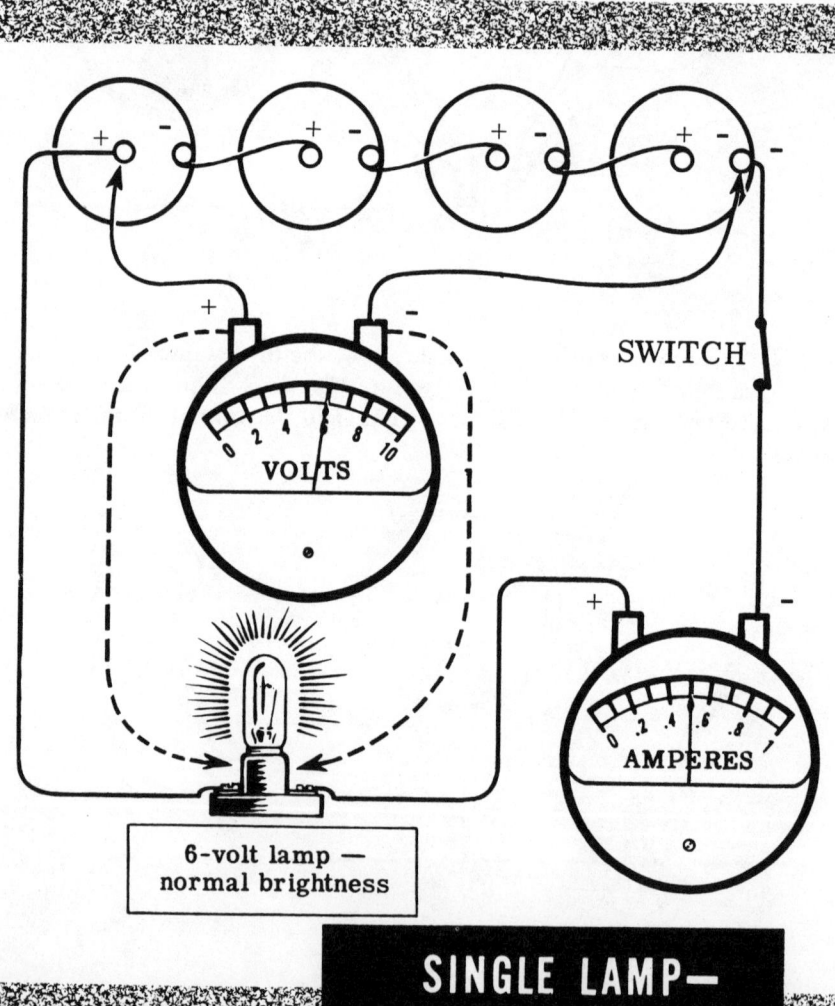

6-volt lamp — normal brightness

SINGLE LAMP— NORMAL CURRENT

Experiment/Application—Series Circuit Resistance (continued)

Next, suppose you replace the single lamp socket by three sockets in series, and again insert 6-volt, 0.5-ampere lamps in the socket. The lamps now light well below normal brilliancy, and the ammeter reading would be about one-third of its previous value. A voltmeter reading taken across the total circuit reads 6 volts, and across each lamp the voltage is 2 volts.

Since the voltage from the battery is not changed, but the current is lower, it follows that the resistance must be greater.

If you measured and added the voltages across each lamp, you would find that the sum of the voltages across the individual resistances (lamps) still equals the total voltage.

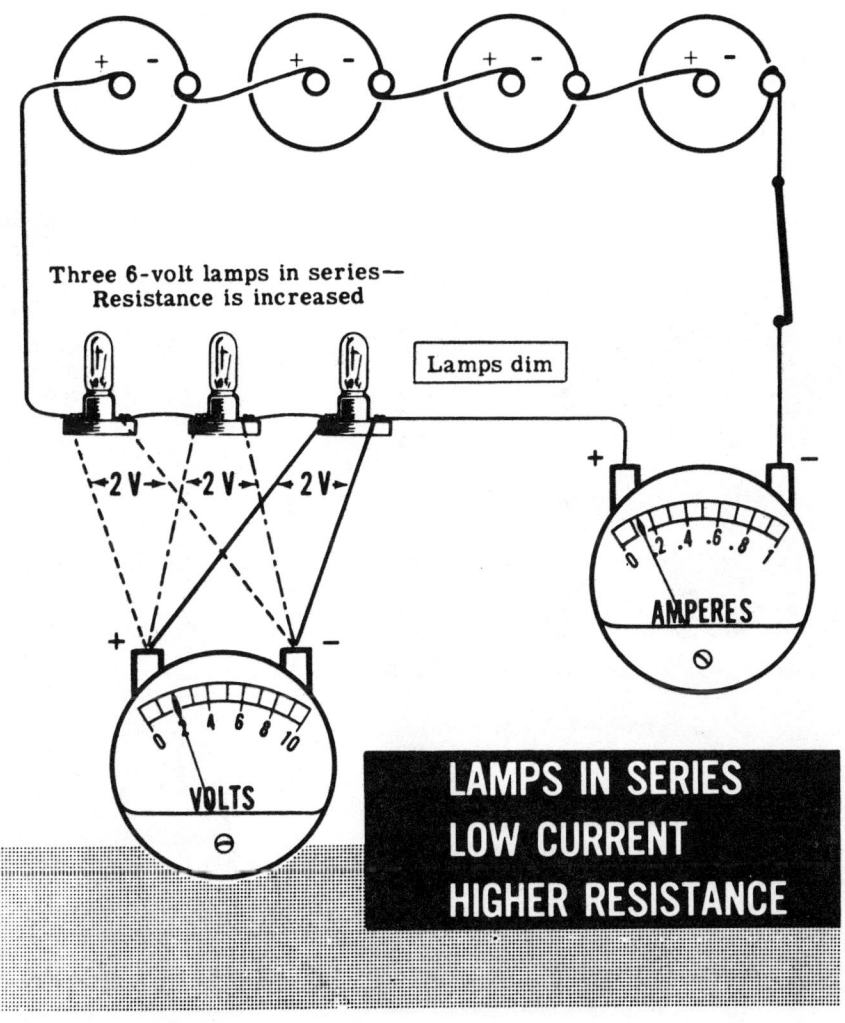

Three 6-volt lamps in series—
Resistance is increased

Lamps dim

**LAMPS IN SERIES
LOW CURRENT
HIGHER RESISTANCE**

DC SERIES CIRCUITS

Experiment/Application—Series Circuit Current

To see the effect of changing resistances on the amount of current flow, and how different pieces of equipment require different amounts of current for proper operation, you could replace one of the 6-volt lamps by a 2.5-volt lamp drawing 0.75 ampere. The two 6-volt lamps will increase in brilliance by almost 50%, while the 2.5-volt lamp will light only dimly. The ammeter reading would show that the current has increased—which proves that decreasing the resistance of one part of the circuit decreases the total opposition to current flow and, hence, *increases* the total circuit current.

Replacement of another 6-volt lamp with a 2.5-volt lamp would further decrease the total resistance and increase the total circuit current.

The brilliance of the lamps increases as the current flow increases; and if the last 6-volt lamp were to be replaced by a lower-resistance 2.5-volt lamp, you would see that the circuit current for the three 2.5-volt lamps is approximately the same as that for a single 6-volt lamp. You would also observe that the three lamps light at about normal brilliance, because the current is only slightly less than the rated value of the lamps, as is the voltage measured across each lamp.

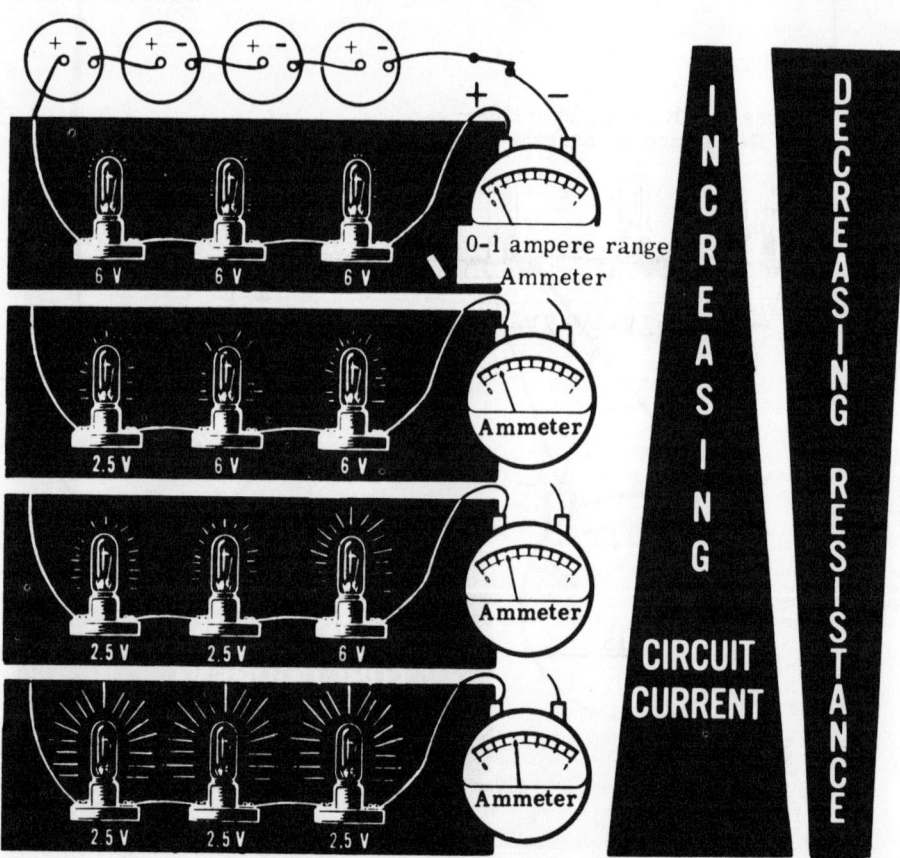

Experiment/Application—Series Circuit Voltage/Kirchhoff's Second Law

The rated voltage of three 2.5-volt lamps in series is 7.5 volts, so that a 6-volt battery does not cause the rated current to flow. Adding one more dry cell would increase the circuit voltage without changing the resistance, thereby causing a greater current flow, as indicated by the increased brilliance of the lamps and the increased current reading. Voltage readings taken across the lamps show that the voltage across each lamp is the rated voltage of 2.5 volts. If you measured the total battery voltage it would also equal 7.5 volts.

Removal of one cell of the battery at a time, followed by taking voltage readings across the lamps, and across the battery, will once again show that the voltages across the lamps are about equal, and always add up to the total battery voltage.

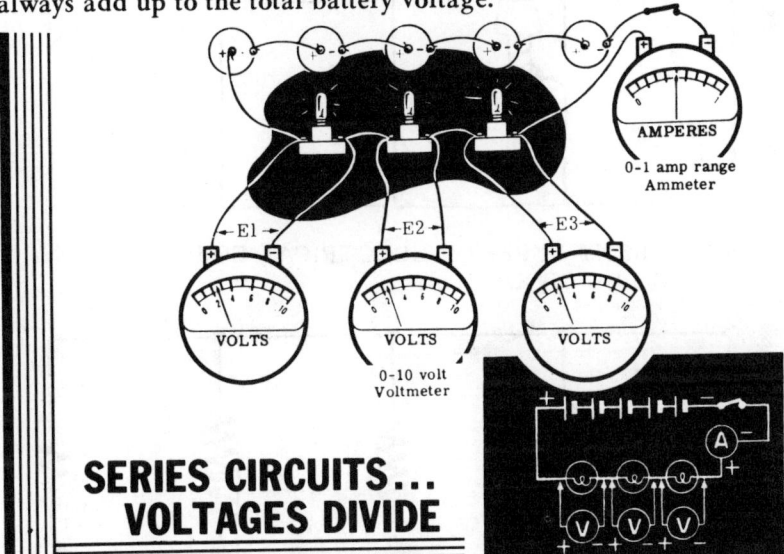

SERIES CIRCUITS... VOLTAGES DIVIDE

With five cells connected to form a 7.5-volt battery, replace one of the 2.5-volt lamps with a 6-volt lamp having greater resistance. Voltmeter readings across the lamps would still total 7.5 volts when added together, but would not all be equal. The voltages across the lower resistance 2.5-volt lamps would be equal but less than 2.5 volts, while the voltage across the higher resistance 6-volt lamp would be greater than 2.5 volts. You can see that for resistors in series the voltage divides in proportion across the various resistances connected in series, with *more* voltage drop across the *larger* resistance and *less* voltage drop across the *smaller* resistance; the total voltage (E_t) is exactly equal to the voltage across each resistance. To put it another way,

$$E_t = E1 + E2 + E3$$

which is *Kirchhoff's Second Law!*

Learning Objectives—Next Section

Overview—Aside from the series circuit that you already can solve, you will learn about *parallel* circuits in the next section. When you can solve both types of circuits, you will find that you can solve any circuit because all circuits are made of combinations of series and/or parallel circuits.

DIFFERENT TYPES OF ELECTRICAL EQUIPMENT IN PARALLEL DIVIDE . . . .

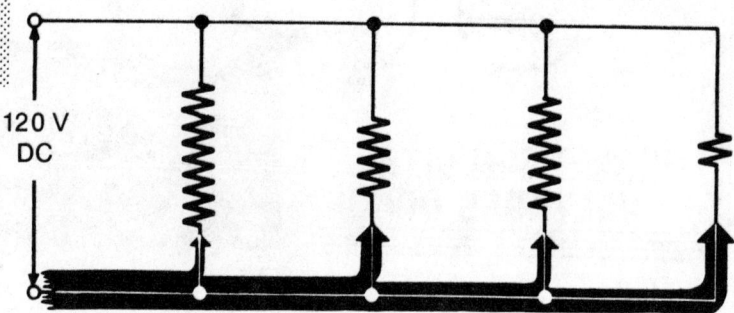

120 V DC

. . . . THE TOTAL CURRENT UNEQUALLY

DC PARALLEL CIRCUITS

The Parallel Circuit

When resistances, instead of being connected *end-to-end* as in a series circuit, are connected *side-by-side* so that there exists more than one path through which current can flow, the resistances are said to be *parallel-connected* or *connected in parallel*; and the circuit of which they form a part is called a *parallel circuit*.

Two lamp sockets, for instance, connected terminal-to-terminal by two pieces of wire, are *parallel-connected*. When any two terminals are connected across a voltage source, the whole arrangement—both lamps, the voltage source, and the wires connecting them together—forms a complete parallel circuit.

In the same way, cells in a battery connected so that there is *more than one path* for current flow through the battery are said to be parallel-connected. Each individual cell in a battery whose cells are so connected furnishes only a *part* of the total current drawn from the battery. As you might suspect, when you connect batteries or cells in parallel, you must connect the *positive* terminals *together* and the *negative* terminals *together*.

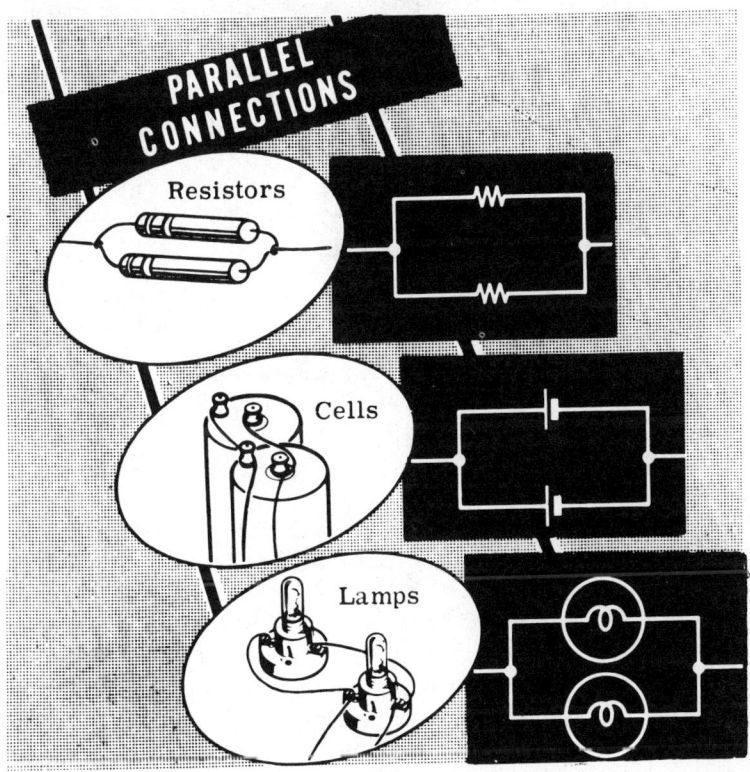

An obvious example of parallel connection is found in the electrical wiring of an ordinary house, in which every one of the various electrical appliances used in the house is connected in parallel across the line.

Voltage in Parallel Circuits

When resistances in parallel are placed across a voltage source, the *voltage* across each of the resistors is always the same. The *current* through each resistor, however, will vary according to the value of the individual resistance.

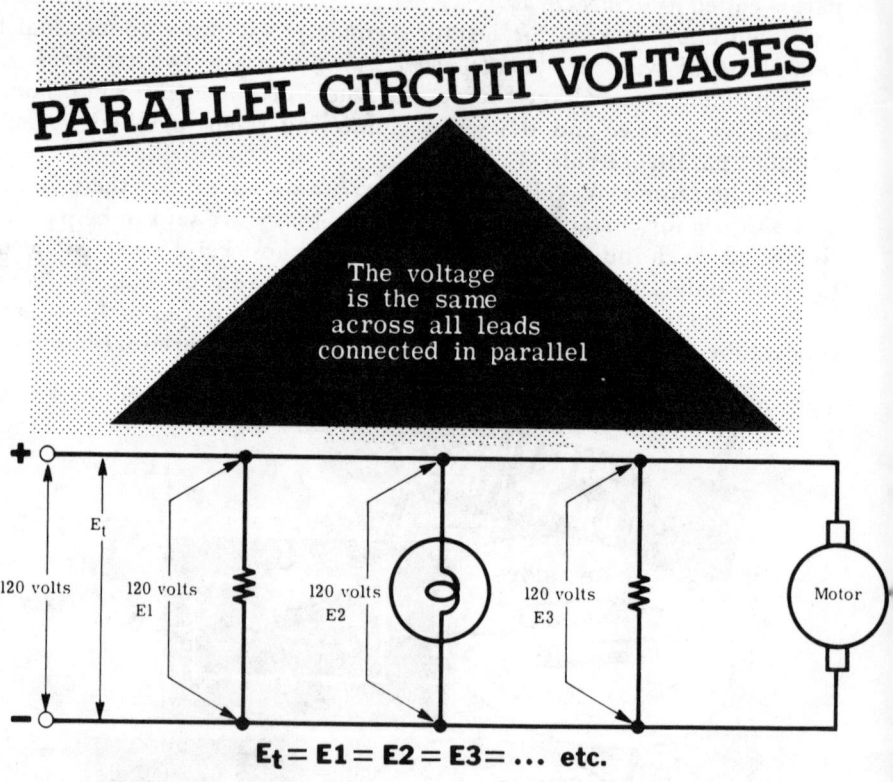

The fact that the voltages applied to each of the resistors or loads in a parallel circuit are always the same has an important practical consequence. All components which are to be connected in parallel must have the *same voltage rating* if they are to work properly.

The line voltage throughout the U.S. is 120 volts. You probably know by now that lamps and other electrical appliances with a voltage rating of 120 volts, or thereabouts, work perfectly well, whereas a lamp-bulb rated at 12 volts burns out immediately because excess current is flowing through it.

The reason is that, since all the appliances are connected across the same voltage source, the same voltage is applied across each. All must, therefore, be of the proper rating to handle this voltage.

DC PARALLEL CIRCUITS

Current Flow in Parallel Circuits

Current flowing through a parallel circuit divides to flow through each of the parallel paths. Take a circuit containing three branches—call them AB, CD, and EF, respectively—connected in parallel. In such a circuit:

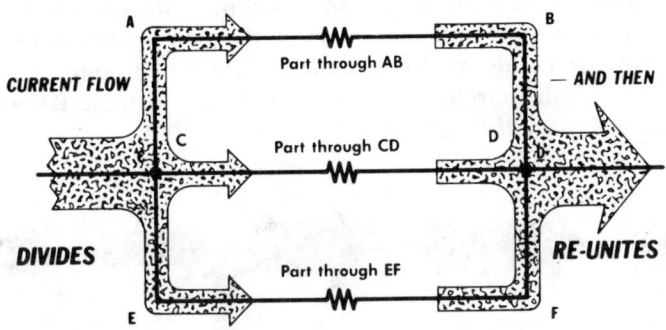

The current flowing through the several branches of a parallel circuit divides in inverse proportions, governed by the comparative resistance of the individual branches. Thus, the lower the resistance of any branch in proportion to the resistance of other branches in the same parallel circuit, the higher will be the proportion of total current flow which that branch will take.

You can put the same thing more simply, and almost as accurately, by saying: *In a parallel circuit, branches having low resistance draw more current than do branches having high resistance.*

In the illustration below, the differing values of the four resistors in the parallel circuit are indicated by the length of the resistor symbol used in each case. Note that the higher the resistance, the smaller the proportion of current flowing through it, and vice versa.

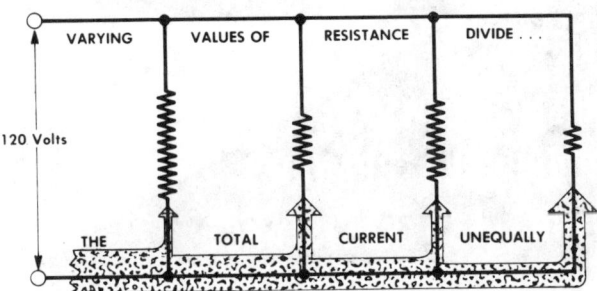

The way in which current divides in a parallel circuit is of great practical importance. For instance, since every electrical appliance used in a house, is (as you know) connected in parallel across the line, current will divide unequally through the differing values of resistance which these appliances present, *the highest current flowing through the lowest resistance.* You will learn how protection against excessive flow is given when you come to read about *fuses* later on.

Current Flow in Parallel Circuits (continued)

It is worth repeating the important facts you have just learned.

When unequal resistances are connected in parallel, opposition to current flow is not the same in every branch of the circuit. A *small* value of resistance offers *less* opposition to current flow. Current flow is always greatest through the path of least opposition; so the smaller resistors in a parallel circuit always pass more current than do the larger ones.

In the circuit below, for instance, a total of 9 amperes is flowing through a parallel circuit consisting of two resistors, R1 and R2, of which R1 has twice the value of R2.

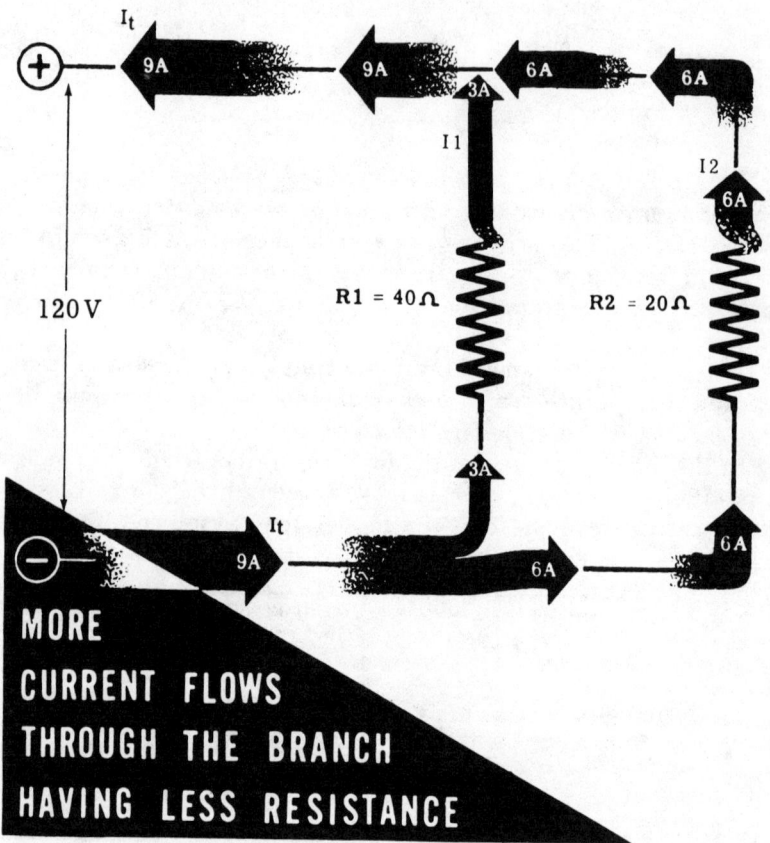

MORE CURRENT FLOWS THROUGH THE BRANCH HAVING LESS RESISTANCE

Note that *the current divides in inverse proportion to the values of the two resistors*—only 3 amperes flow through the 40-ohm resistance of R1, while 6 amperes flow through the 20-ohm resistance of R2. If the value of R1 were to be quadrupled to 160 ohms, the current through R1 would be reduced from 3 amperes to 3/4 ampere while the current through R2 would remain *unchanged*. Thus, the total current would be 6 amperes + 3/4 ampere = 6.75 amperes.

Current Flow in Parallel Circuits (continued)

You may have noticed on the previous page that the total current in a parallel circuit seems to be equal to the sum of the current in each component. This is always true, and can be expressed as

$$I_t = I_1 + I_2 + I_3 + \ldots$$

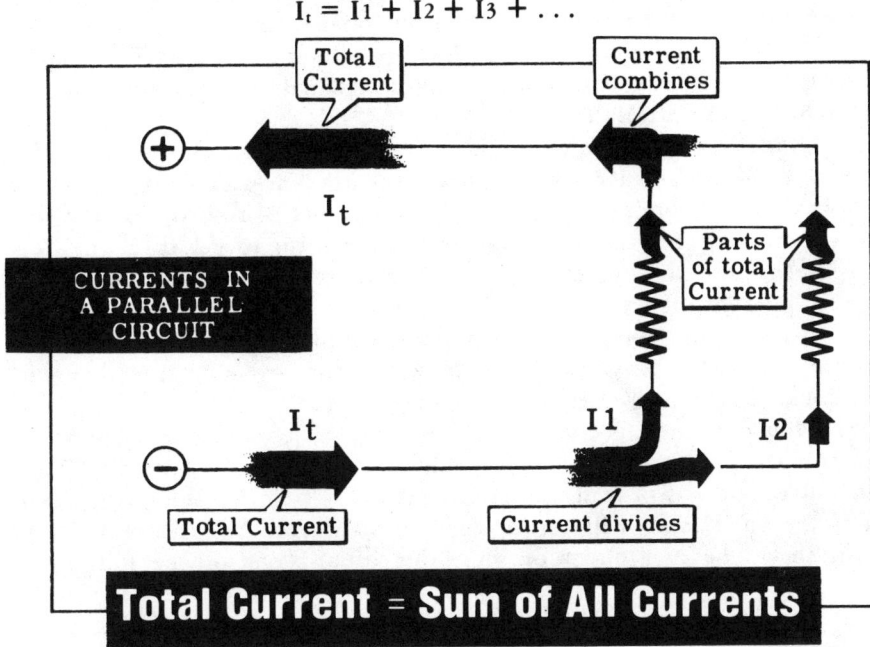

When a circuit consists of *equal* resistances in parallel, the current flowing through each and all of the resistances will be equal; and the current, seeking always for the line of least resistance, will find equal opposition in every path it can take.

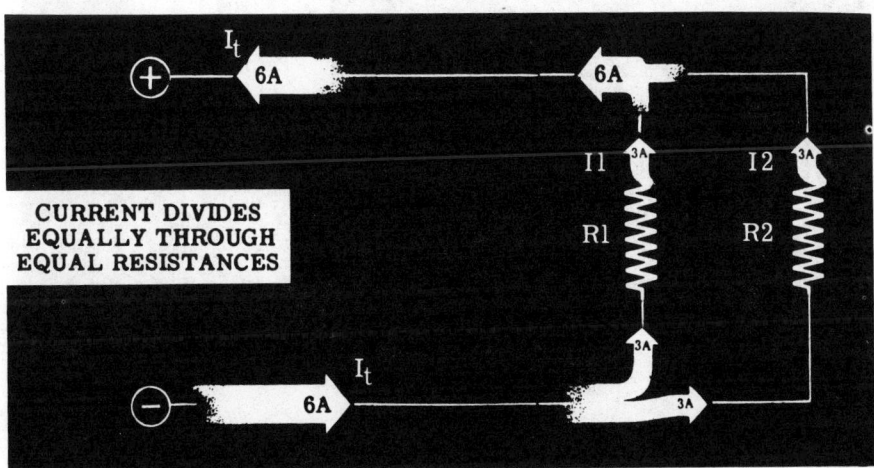

Equal Resistors in Parallel Circuits

In a circuit such as the one illustrated at the bottom of the preceding page, the current flow through each of the two resistors is plainly equal to the total current divided by the number of resistors connected in parallel. This rule will always apply no matter how many resistors there may be, provided that every resistor is of equal value.

The next step is to determine how great is the *total effective resistance* which a circuit, consisting of a number of equal resistors, will offer to current flowing through it.

You know that one of the factors which diminishes the value of a given resistor is an increase in its cross-sectional area. This is exactly the effect which is obtained by connecting a number of resistors in parallel. There is a bigger cross-sectional area for current to get through, so it passes through the parallel combination of resistors *more easily* than it would pass through a single one of them.

It is just like having two water pipes laid side-by-side in a system. Given a constant pressure (head) of water, the two pipes together will obviously pass more water than either could alone.

The conclusion is that *resistors or loads connected in parallel present a lower combined resistance or load than does any one of them individually.* If your load consists of four 200-ohm resistors in parallel, the resistance of the combined load will (since current divides equally through equal resistances) be one-fourth of the value of any one of the individual resistors.

Thus, the total load will be 50 ohms; and a parallel connection, such as that shown below, will act in a circuit as if it were *equivalent* to a single resistance of 50 ohms. It will often be convenient to combine parallel loads into a single one for purposes of calculation.

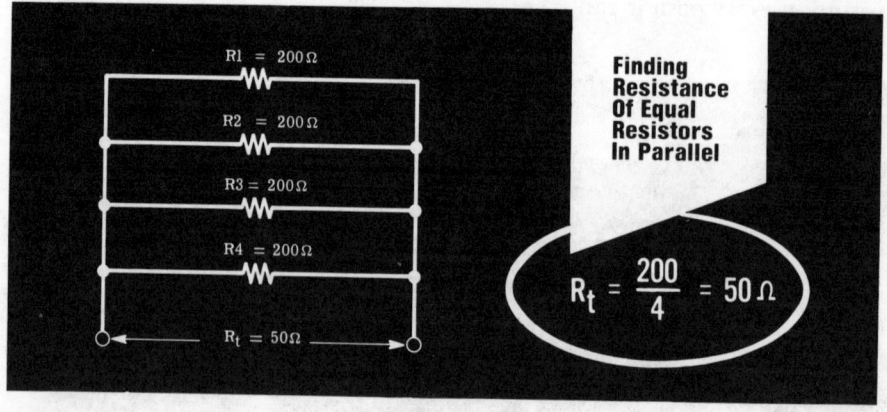

Finding Resistance Of Equal Resistors In Parallel

$$R_t = \frac{200}{4} = 50 \, \Omega$$

Unequal Resistors in Parallel Circuits

When a circuit contains resistors in parallel whose values are *unequal*, the problem of assessing total resistance becomes more difficult.

In favorable circumstances, you can find out the equivalent value of resistance offered by two unequal resistors in parallel by using an ohmmeter. In a parallel circuit consisting of two resistors, R1 and R2, whose values are 60 and 40 ohms, respectively, you would, in fact, get an ohmmeter reading of 24 ohms for total resistance.

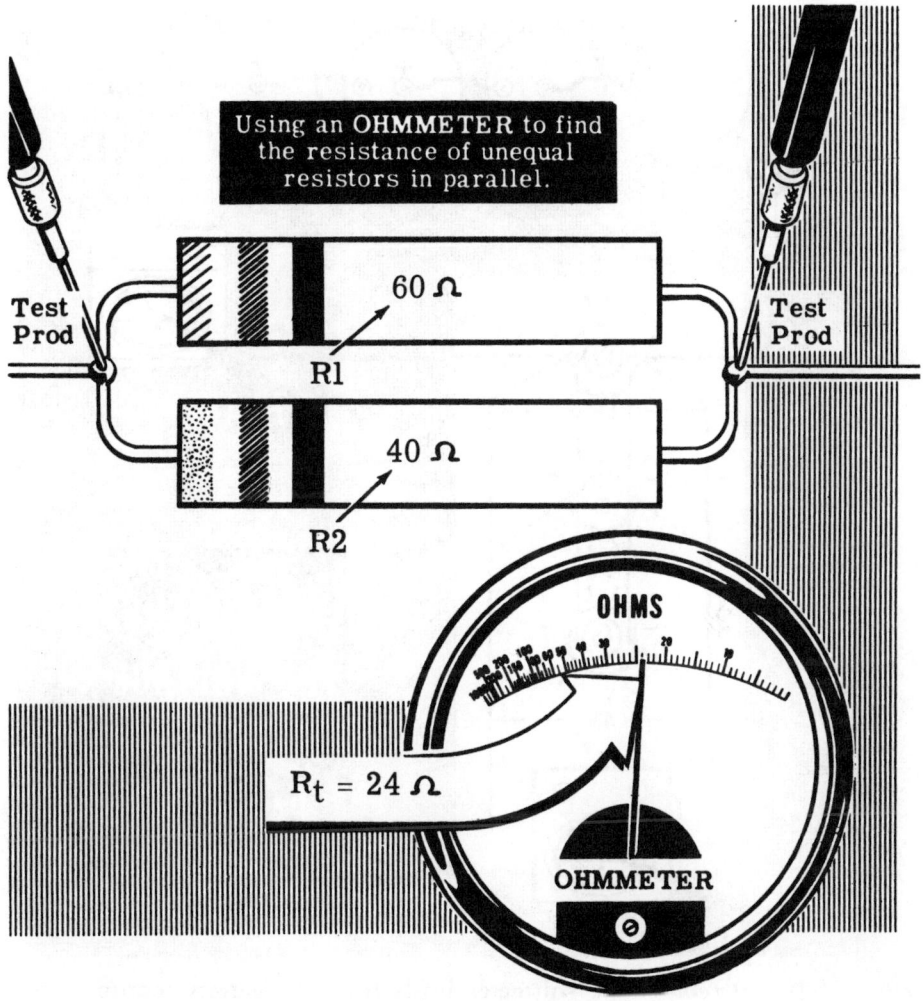

Using an OHMMETER to find the resistance of unequal resistors in parallel.

And now for some experiments to demonstrate further and apply your understanding of dc parallel circuits.

Experiment/Application—Parallel Circuit Voltage

While the current through the various branches of a parallel circuit is not always the same, the voltage across each branch resistance is equal to that across the others. If you connect three lamp sockets in parallel and insert 6-volt, 250-mA (0.25 ampere) lamps in each of the three parallel-connected sockets, you will see that each lamp lights with the same brilliance as when only a single lamp is used, and that for three lamps the circuit current is 750 mA. Also, you will see that the voltmeter reading across the battery terminals is the same whether one, two, or three lamps are used.

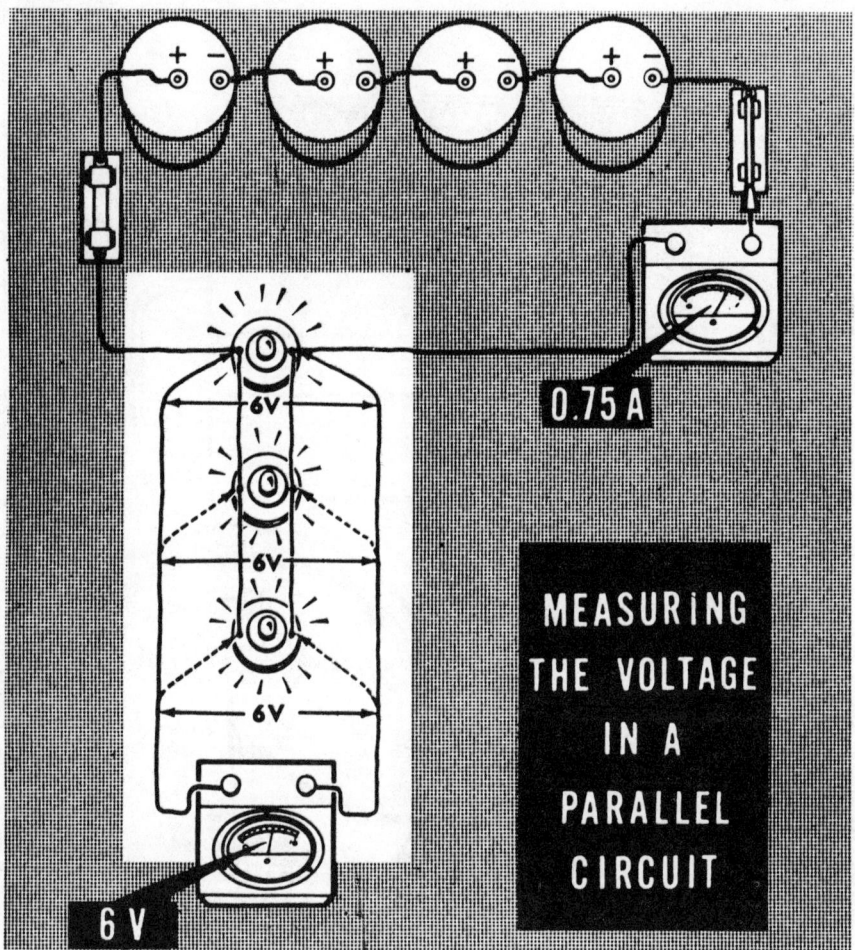

MEASURING THE VOLTAGE IN A PARALLEL CIRCUIT

If you remove the voltmeter leads from the battery terminals, and connect the voltmeter across the terminals of each lamp socket in turn, you see that the voltage is the same across each of the lamps and is equal to that of the voltage source—the battery. This demonstrates that the voltage across each circuit element is the same in a parallel circuit.

Experiment/Application—Parallel Circuit Current

To demonstrate the division of current, you could replace a 250-mA lamp with a 150-mA lamp, as shown. The ammeter now shows that the total circuit current is 400 mA. By connecting the ammeter first in series with one lamp, then in series with the other, you will see that the 400-mA total current divides, with 250 mA flowing through one lamp and 150 mA through the other. Then connect the ammeter to read the total circuit current at that end of the parallel combination opposite the end at which it was originally connected. You will see that the total circuit current is the same at each end of the parallel circuit, the current dividing to flow through the parallel branches of the circuit and combining again after passing through these branches.

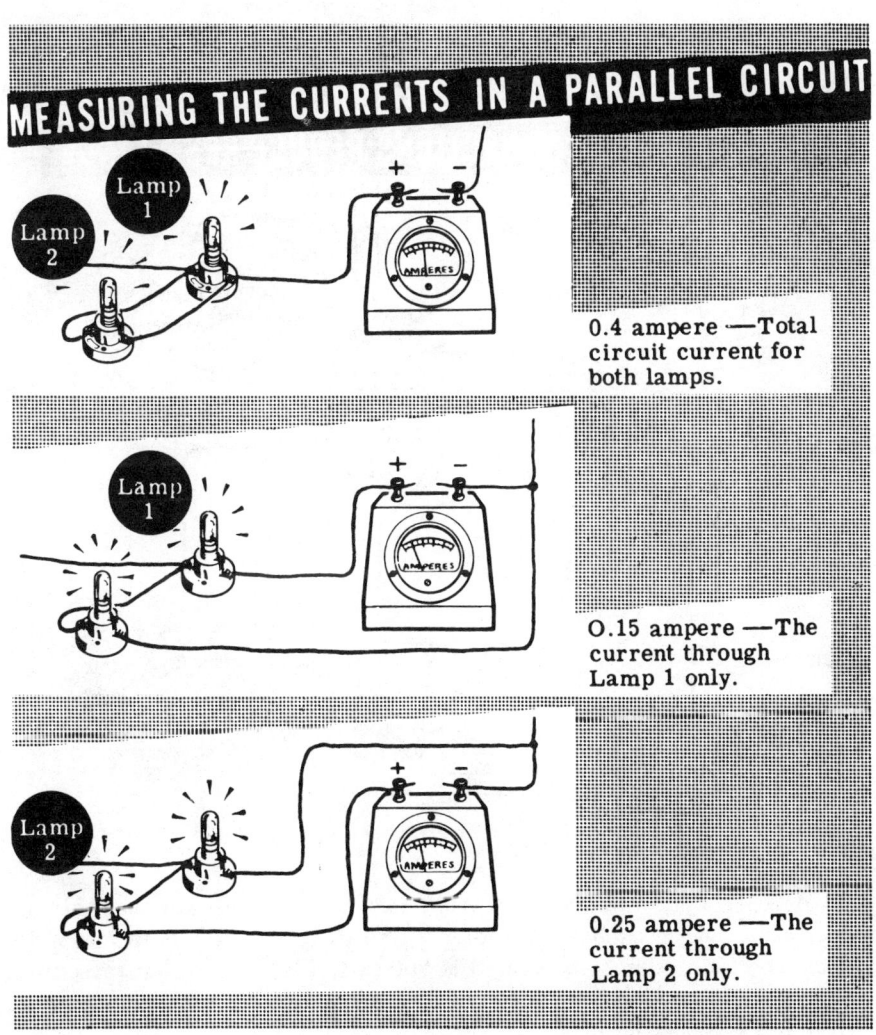

Experiment/Application—Parallel Circuit Resistance

When the total current flow in a circuit increases with no change in the voltage, a decrease in total resistance is indicated. To show this effect, you could connect two lamp sockets in parallel. This parallel combination is connected across the terminals of a 6-volt battery of dry cells with an ammeter inserted in one battery lead to measure the total circuit current. As a voltmeter is connected across the battery terminals you will see that the voltage is 6 volts. When only one 250-mA, 6-volt lamp is inserted, the ammeter indicates a current flow of approximately 250 mA, and the voltmeter reads 6 volts. With both lamps inserted, the current reading increases to 500 mA (0.5 ampere), but the voltage remains at 6 volts—indicating that the parallel circuit offers less resistance than a single lamp.

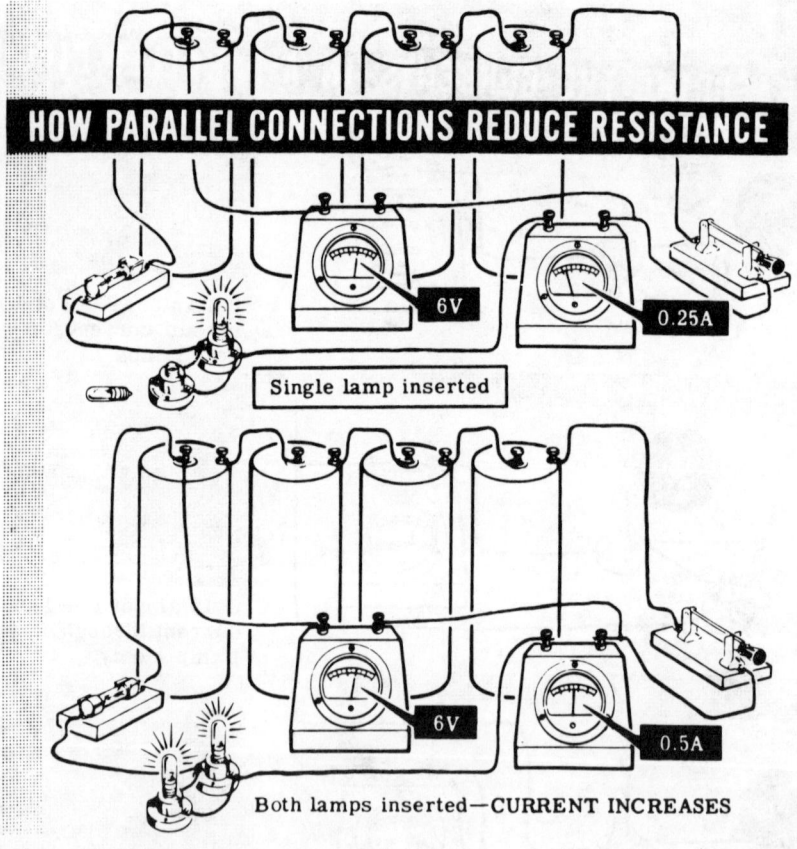

HOW PARALLEL CONNECTIONS REDUCE RESISTANCE

Single lamp inserted

Both lamps inserted—CURRENT INCREASES

As each of the lamps is inserted in turn, you will see that the ammeter reading for each lamp alone is 250 mA; but with both lamps inserted, the total current indicated is 500 mA. This shows that the circuit current of 500 mA divides into two 250-mA currents, with each flowing through a separate lamp, or path.

Experiment/Application—Parallel Resistances

To show how parallel connection of resistances decreases the total resistance, you could measure the resistance of three 330-ohm resistors individually with the ohmmeter. When two of the 330-ohm resistors are paralleled, the total resistance should be 165 ohms; this is shown by connecting them and measuring the parallel resistance with the ohmmeter. As another 330-ohm resistor is connected in parallel, you will see that the resistance is lowered to a value of 110 ohms. This not only shows that connecting *equal* resistances in parallel reduces the total resistance, but also that the total resistance can be found by dividing the value of a single resistance by the number of resistances used.

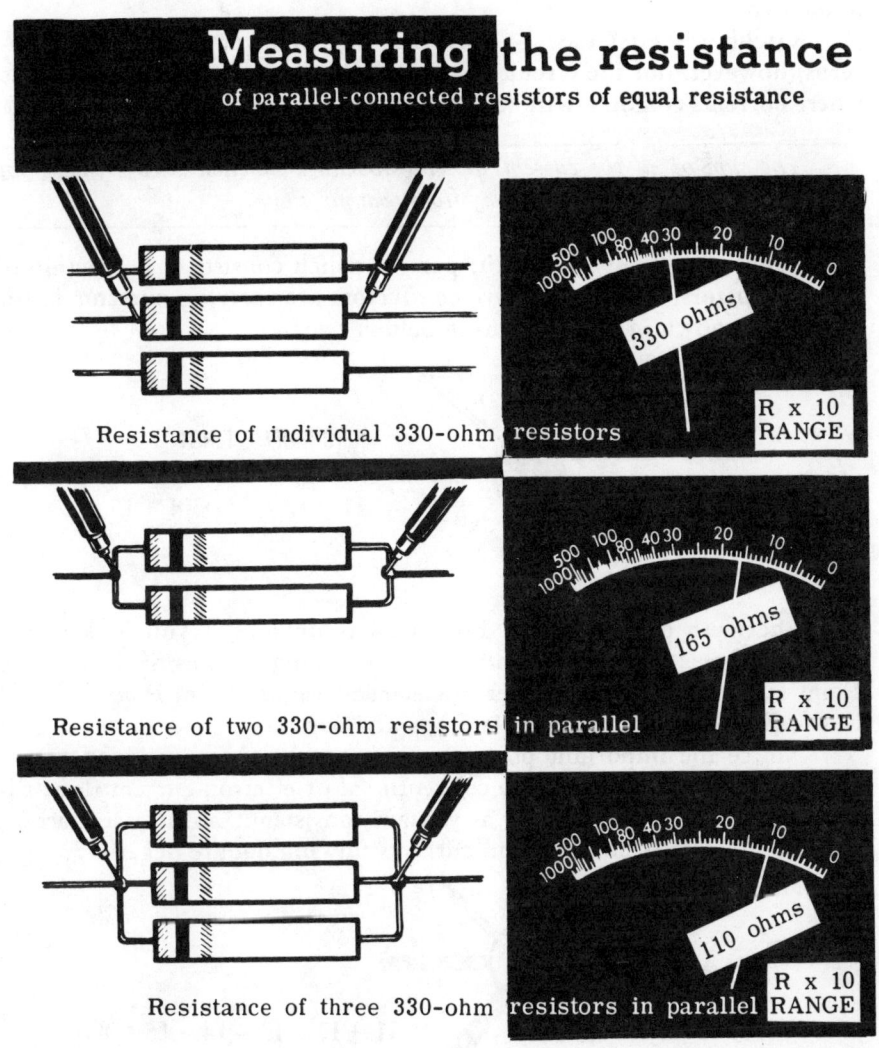

Measuring the resistance of parallel-connected resistors of equal resistance

Resistance of individual 330-ohm resistors — 330 ohms (R x 10 RANGE)

Resistance of two 330-ohm resistors in parallel — 165 ohms (R x 10 RANGE)

Resistance of three 330-ohm resistors in parallel — 110 ohms (R x 10 RANGE)

Kirchhoff's First Law

It is often not possible to use an ohmmeter to get resistance readings—and you ought, in any case, to know how such answers can be found by calculation. You will not be surprised to hear that it is quite easily done by intelligent use of Ohm's Law—but you also will need help from another equation which you have been waiting for some time to hear about. It is known as *Kirchhoff's First Law*.

You already know that in a series circuit the current entering the circuit is exactly equal to the current leaving the circuit. If you have grasped correctly the idea of current flow as a stream of electrons traveling around their circuit, you will see at once that the statement above must be true whether the circuit around which the current is flowing is a series circuit, a parallel circuit, or a circuit containing any combination of the two.

Kirchhoff's First Law is, thus, true of every type of circuit. It concerns, however, not the circuit as a whole but only individual junctions where currents combine within the circuit itself. It states:

> *The sum of all the currents flowing toward a junction always equals the sum of all the currents flowing away from that junction.*

Suppose you have a circuit, part of which consists of a junction of five conductors, and that all five conductors are carrying currents in the directions shown in the illustration below.

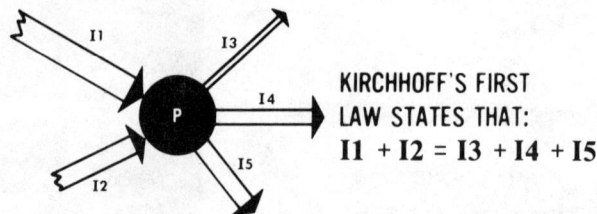

KIRCHHOFF'S FIRST LAW STATES THAT:
$I_1 + I_2 = I_3 + I_4 + I_5$

The truth of Kirchhoff's First Law is obvious if you look at the drawing above. Currents I_1 and I_2 are delivering streams of electrons to Point P. Therefore, the number of electrons leaving Point P *must* always be the same as the number of electrons arriving there.

Notice the important point that *direction* has been assigned to the current flow. Whether you use conventional or electron-current flow, direction is unimportant as long as you are consistent. In this case, current into the junction is positive, and currents flowing out are negative.

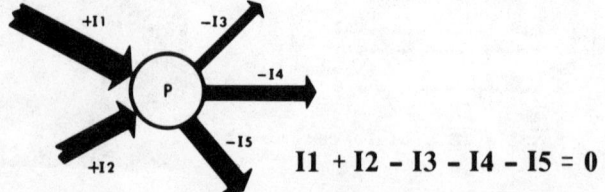

$I_1 + I_2 - I_3 - I_4 - I_5 = 0$

DC PARALLEL CIRCUITS

Kirchhoff's First Law (continued)

To use Kirchhoff's First Law in a complete circuit, the rule is (as always): *First draw the circuit*. Then indicate on the circuit diagram the direction of current flow through every resistance in the circuit. Then determine which of these currents flows toward, and which away from, every junction in the circuit. Mark this information in on the circuit diagram. The value and direction of flow of *unknown* currents can then often be determined by applying the Law.

At the circuit junction pictured below, neither the direction nor the value of I_1 is known.

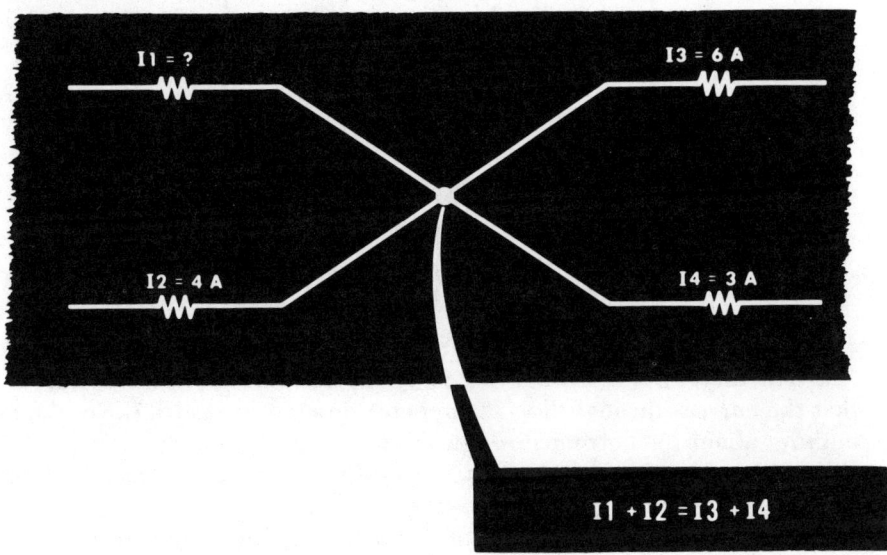

$$I_1 + I_2 = I_3 + I_4$$

The direction of the unknown current is first determined by adding together all the known currents flowing toward the junction and all those flowing away from it, and then comparing the two. Here, the sum of the currents flowing away from the junction ($I_3 = 6$ amperes plus $I_4 = 3$ amperes, making a total of 9 amperes) is greater than is the value of the one known current ($I_2 = 4$ amperes) flowing into the junction.

It follows that the unknown current (I_1) must also be flowing *into* the junction, or else the balance of electron flow at the junction could not be maintained. Its value can be determined by substituting known values in the Kirchhoff equation, $I_1 + I_2 = I_3 + I_4$.

Therefore $I_1 + 4\,A = 6\,A + 3\,A = 9\,A$
$I_1 = 9\,A - 4\,A = 5\,A$

Also using $I_1 + I_2 - I_3 - I_4 = 0$
$I_1 + 4\,A - 6\,A - 3\,A = 0$
$I_1 - 5\,A = 0$
$I_1 = 5\,A$

Kirchhoff's First Law (continued)

Now take a slightly more complicated example of using Kirchhoff's First Law to find out the values and directions of flow of unknown currents in a circuit.

Suppose you have a circuit consisting of seven resistors, connected as shown in the diagram below. This is a series-parallel circuit, a type about which you will learn much more later.

THIS IS HOW YOUR CIRCUIT LOOKS

[Circuit diagram showing resistors R1 through R7 connected in a series-parallel configuration. R1 on the left connects to a network where R3 is on top-left, R6 is on top-right, R2 is on bottom-left, R5 is on bottom-right, R4 is in the middle vertical branch, and R7 is on the right.]

You know that the current through R2 is 7 amperes flowing toward R5; that the current through R3 is 3 amperes flowing toward R6; and that the current through R5 is 5 amperes flowing toward R7. You know nothing about the current through the resistors R1, R4, R6, and R7; but you need to know both their values and the directions in which they are flowing. Here is how it is done.

Draw the circuit in symbolic form, designating all currents, with values and directions, if known. Then identify each junction of two or more resistors with a letter.

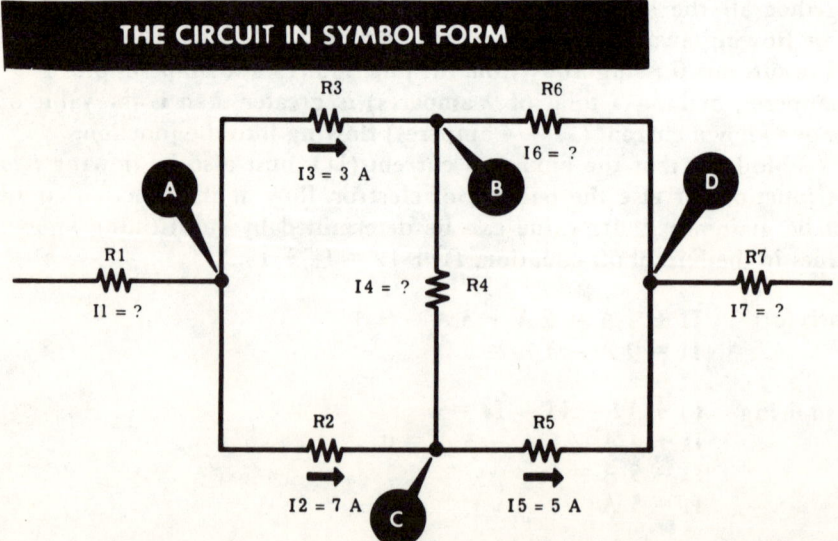

THE CIRCUIT IN SYMBOL FORM

Kirchhoff's First Law (continued)

Find the unknown currents at all junctions where only one current is unknown; then you can use these new values to find unknown values at other junctions.

From the circuit you can see that junctions A and C have only one unknown. So start by finding the unknown current at junction A:

Of the three currents at junction A—I_1, I_2, and I_3—both I_2 and I_3 are known, and flow away from the junction. I_1 must, therefore, flow toward the junction, and its value must be equal to the sum of I_2 and I_3.

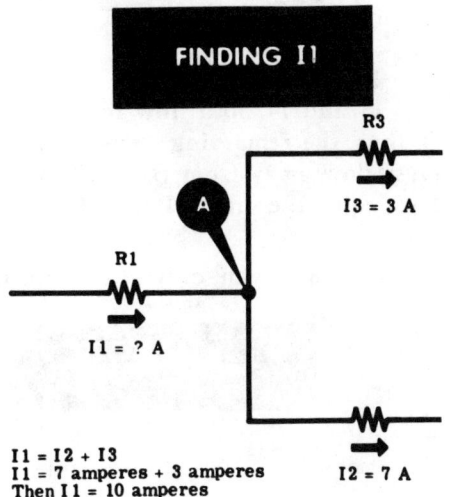

$I_1 = I_2 + I_3$
$I_1 = 7$ amperes + 3 amperes
Then $I_1 = 10$ amperes

Next find the unknown current at junction C:

At C two currents—I_2 and I_5—are known, and only I_4 is unknown. Since I_2, flowing toward C, is greater than I_5, flowing away from C, then the third current I_4 also must flow away from C. Also, since the current flowing toward C equals that flowing away from it, it follows that I_2 equals I_4 plus I_5.

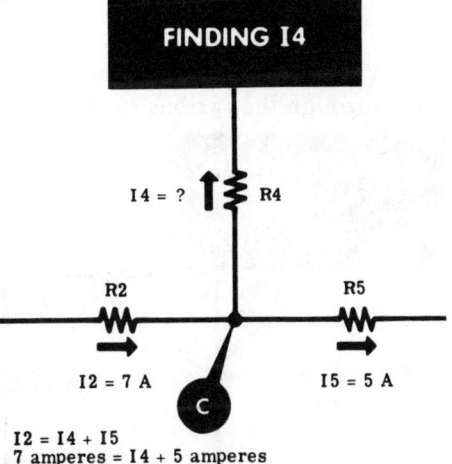

$I_2 = I_4 + I_5$
7 amperes = I_4 + 5 amperes
Then $I_4 = 2$ amperes

Kirchhoff's First Law (continued)

Now that the value and direction of I4 are known, only I6 is unknown for junction B. You can find the amount and direction of I6 by applying the law for current at B.

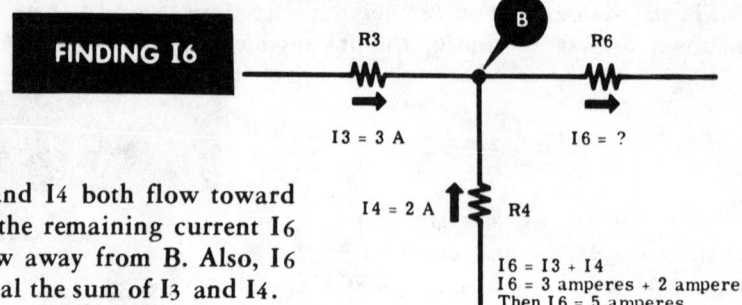

FINDING I6

I3 and I4 both flow toward B; thus, the remaining current I6 must flow away from B. Also, I6 must equal the sum of I3 and I4.

I6 = I3 + I4
I6 = 3 amperes + 2 amperes
Then I6 = 5 amperes

With I6 known, only I7 remains unknown, at junction D.

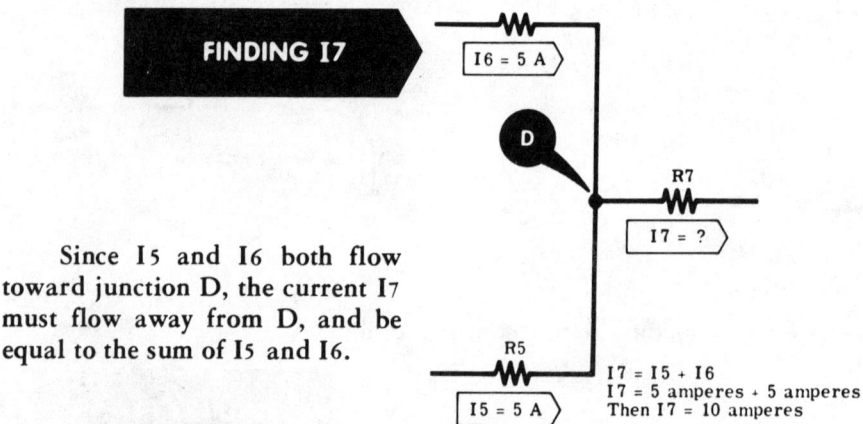

FINDING I7

Since I5 and I6 both flow toward junction D, the current I7 must flow away from D, and be equal to the sum of I5 and I6.

I7 = I5 + I6
I7 = 5 amperes + 5 amperes
Then I7 = 10 amperes

You now know all of the circuit currents and the directions of their flow through the various resistors.

CIRCUITS WITH ALL THE CURRENTS KNOWN

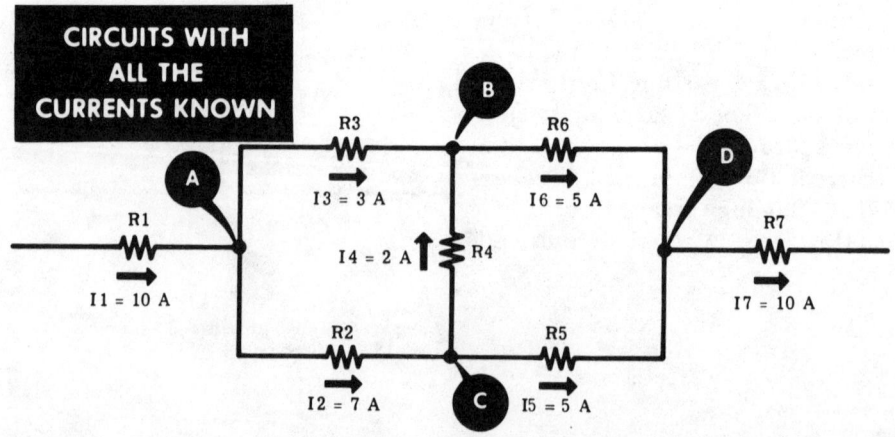

Experiment/Application— Kirchhoff's First Law

Check for yourself Kirchhoff's First Law—the law of circuit currents. Suppose you connect a 15-ohm resistor in series with a parallel combination of three 15-ohm resistors, and then connect the entire circuit across a 9-volt dry cell battery with a switch and fuse in series.

The total resistance of the circuit is 20 ohms, resulting (by Ohm's Law) in a total circuit current of 0.45 ampere. This total current must flow through the circuit from the negative (−) to positive (+) battery terminals (see the circuit diagram below). At junction (a), the circuit current—0.45 ampere—divides to flow through the three parallel resistors toward junction (b). Since the parallel resistors are all equal, the current divides equally, with 0.15 ampere flowing through each resistor. At junction (b), the three parallel currents combine, to flow away from the junction through the series resistor.

If you connect an ammeter to read the current in each lead at junction (b), you will see that the sum of the three currents flowing toward the junction is equal to the current flowing away from the junction.

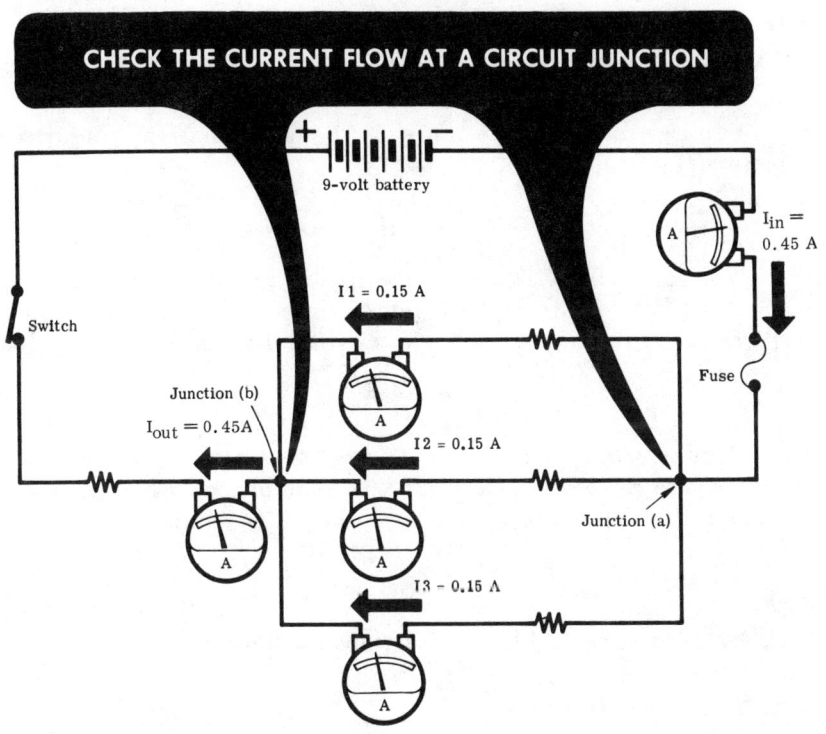

$$\text{Current flowing toward the junction} = \text{Current flowing away from the junction}$$

$$I_{in} = I_1 + I_2 + I_3$$
$$I_1 + I_2 + I_3 = I_{out}$$
$$0.15 + 0.15 + 0.15 = 0.45 \text{ ampere}$$

Unequal Resistors in Parallel Circuits (continued)

With this understanding of Kirchhoff's First Law you are now ready to return to the subject of unequal resistors in parallel circuits and to find out how to calculate the effective resistance of any number of unequal resistors so connected in parallel.

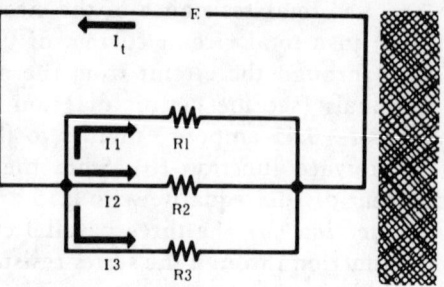

In the circuit opposite, the current divides to pass through three resistors. Apply Kirchhoff's First Law and Ohm's Law to the circuit exactly as you did before.

You know from Kirchhoff's First Law that, in a circuit like the one above,

(1) $I_t = I_1 + I_2 + I_3$

Also, from Ohm's Law you know that

(2) $I_t = \dfrac{E}{R_t}$, $I_1 = \dfrac{E}{R1}$, $I_2 = \dfrac{E}{R2}$, $I_3 = \dfrac{E}{R3}$

You can substitute the values in the four equations (2) for those in equation (1), which gives

(3) $\dfrac{E}{R_t} = \dfrac{E}{R1} + \dfrac{E}{R2} + \dfrac{E}{R3}$

Now use a little algebra on equation (3), and remember that you can do almost anything to one side of an equation provided you do exactly the same to the other side. So, dividing both sides by E yields

$$\boxed{\dfrac{1}{R_t} = \dfrac{1}{R1} + \dfrac{1}{R2} + \dfrac{1}{R3}}$$

Apply to this formula the values assumed for the parallel circuit above where R1 = 300 ohms, R2 = 200 ohms, and R3 = 60 ohms. Substitute these values in the formula you have just found, and you get

$$\boxed{\dfrac{1}{R_t} = \dfrac{1}{300} + \dfrac{1}{200} + \dfrac{1}{60}}$$

Before you can add these fractions together, you must give them the same denominator. In this case, the lowest common denominator is 600; and your equation becomes

$$\boxed{\dfrac{1}{R_t} = \dfrac{2}{600} + \dfrac{3}{600} + \dfrac{10}{600} = \dfrac{15}{600}}$$

Now turn (invert) both sides of the equation ($\dfrac{1}{R_t} = \dfrac{15}{600}$) upside down, and you get

$$\boxed{R_t = \dfrac{600}{15} = 40 \text{ ohms}}$$

Unequal Resistors in Parallel Circuits (continued)

The formula you have just worked out can easily be extended to take account of any number of resistors in a parallel combination. If you had five resistors in your parallel circuit, for instance, the formula for their composite resistance would be

$$\frac{1}{R_t} = \frac{1}{R1} + \frac{1}{R2} + \frac{1}{R3} + \frac{1}{R4} + \frac{1}{R5}$$

A simplified formula for two resistors of unequal value can be derived by extending the formula from the previous page for any number of resistors in parallel.

Take a simple circuit like the one opposite, with a current flowing through a pair of resistors connected in parallel, and then back to the voltage source. You know that

$$\frac{1}{R_t} = \frac{1}{R1} + \frac{1}{R2}$$

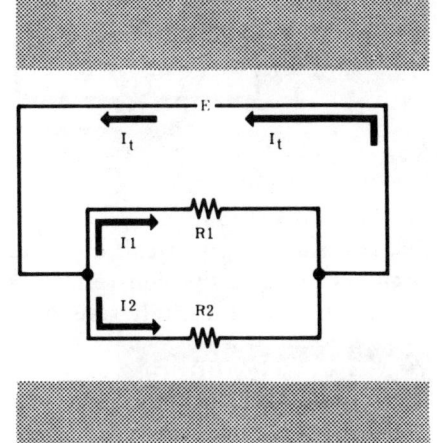

A simple mathematical formula for finding the least common denominator is

$$\frac{1}{A} + \frac{1}{B} = \frac{B+A}{AB}$$

Use of this rule in the equation for two parallel resistors yields

$$\frac{1}{R_t} = \frac{R2+R1}{R1 \times R2}$$

Since we have found a least common denominator, we can turn both sides of the third equation upside down (invert), and get

$$R_t = \frac{R1 \times R2}{R1+R2}$$

You can express the above formula in words as follows: *Total resistance in a parallel circuit composed of two resistors of unequal value is found by MULTIPLYING the value of one resistor by the value of the second; by ADDING the value of one resistor to the value of the second; and then by DIVIDING the first result by the second result.* You will find this a very useful equation to know.

Unequal Resistors in Parallel Circuits (continued)

You can now apply this formula to the parallel circuit containing two resistors with values of 60 and 40 ohms which you were considering on page 2-65. Recall that an ohmmeter reading of the total resistance of such a circuit would be 24 ohms. Let us see if the formula gives the same result.

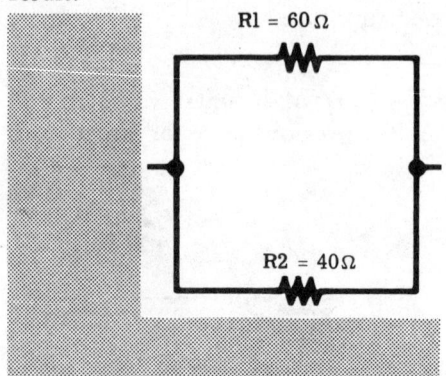

1. *Multiply* the values of the two resistors:

   $60 \times 40 = 2,400$

2. *Add* the values of the two resistors:

   $60 + 40 = 100$

3. *Divide* the product by the sum:

   $\dfrac{2,400}{100} = 24$ ohms

So you see that the *calculated* value of the total resistance and its *indicated* value are identical; and you can confidently say that the parallel combination of a 60-ohm resistor and a 40-ohm resistor will always act as though it were a single resistor with a value of 24 ohms.

Another Example

This formula is so useful that it is worthwhile working through one more example to fix it firmly in your mind.

When two resistors, R1 = 120 ohms and R2 = 60 ohms, are connected in parallel, what is their total resistance?

Draw the circuit diagram, and write down the formula:

$R_t = \dfrac{R1 \times R2}{R1 + R2}$

Substitute known values in the formula:

$R_t = \dfrac{120 \times 60}{120 + 60}$

$= \dfrac{7,200}{180}$

$= 40$ ohms

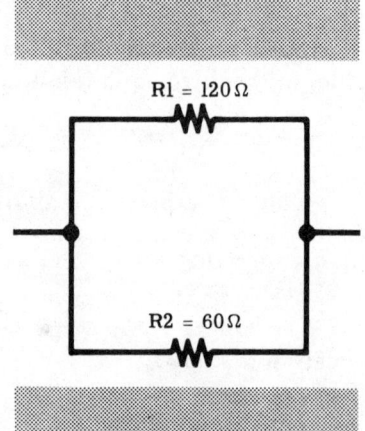

The total resistance of the parallel combination is, thus, 40 ohms, and the combination will act as if it were a single resistor of that value.

Unequal Resistors in Parallel Circuits (continued)

You can use the formula for finding the value of two resistors in parallel to solve more complicated problems by successive application of the formula to reduce the circuit to simpler terms.

Suppose you have three resistors connected in parallel—R1 with a value of 300 ohms, R2 with a value of 200 ohms, and R3 with a value of 60 ohms. What is the effective resistance of the combination?

Take, first, R1 and R2 alone. Substitute their known values in the formula $R_t = \frac{R1 \times R2}{R1+R2}$ and you will get a value for the *equivalent resistance* of the R1 and R2 combination, which we will call R_a. Then combine R_a with R3 in the same way and your answer will be the effective resistance of the complete circuit.

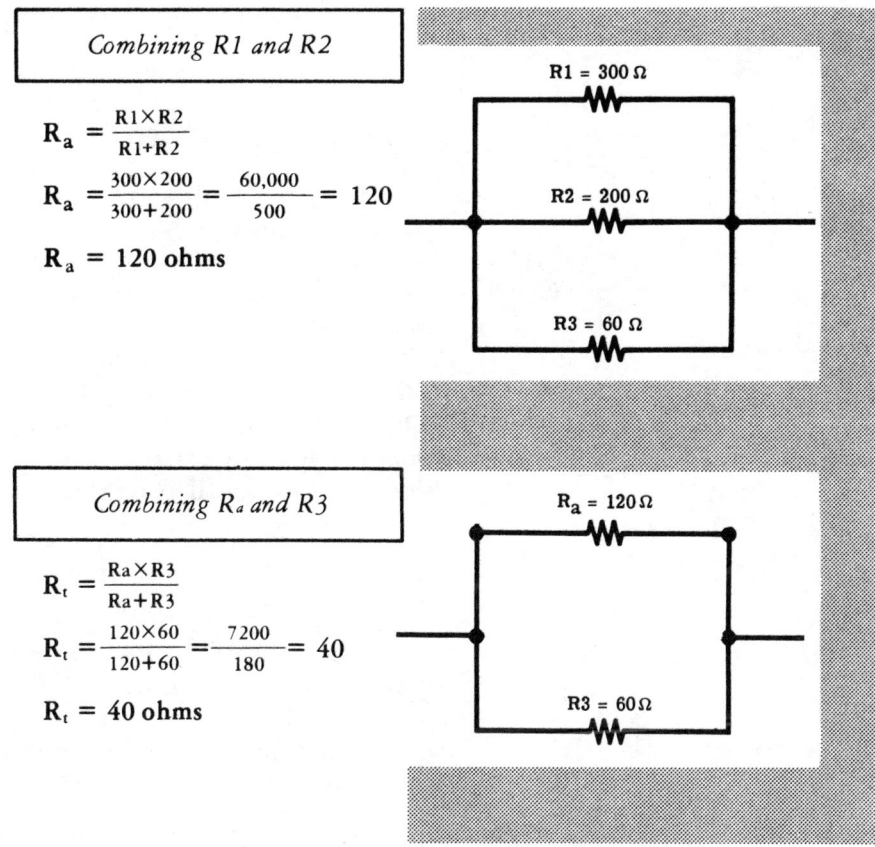

Combining R1 and R2

$R_a = \frac{R1 \times R2}{R1+R2}$

$R_a = \frac{300 \times 200}{300+200} = \frac{60,000}{500} = 120$

$R_a = 120$ ohms

Combining R_a and R3

$R_t = \frac{Ra \times R3}{Ra+R3}$

$R_t = \frac{120 \times 60}{120+60} = \frac{7200}{180} = 40$

$R_t = 40$ ohms

The total resistance of the three resistors connected in parallel is 40 ohms, and the combination will act as a single 40-ohm resistor in the circuit.

You will note that this is the same answer that you got on page 2-76 by applying the basic formula for resistance in parallel.

Review of Parallel Circuits

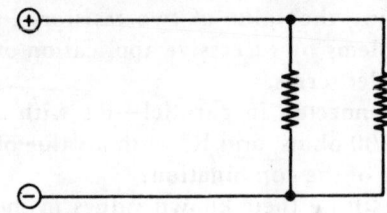

1. **PARALLEL CIRCUIT** — The circuit formed when resistances are connected side-by-side across a voltage source.

2. **PARALLEL CIRCUIT RESISTANCE** — The total resistance in a parallel circuit is lower than is that of the smallest individual resistance in the circuit.

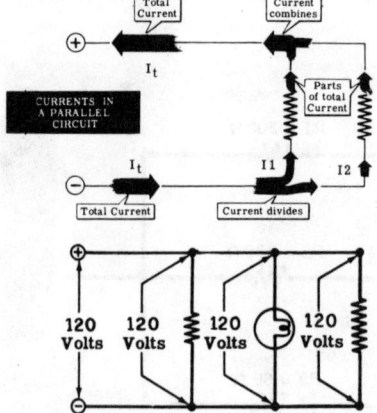

3. **PARALLEL CIRCUIT CURRENT** — The current divides to flow through the parallel branches of the circuit—equally if all the resistors in the circuit are of equal value; unequally, if they are not.

4. **PARALLEL CIRCUIT VOLTAGE** — The voltage across every resistance in a parallel circuit is the same, and is equal to that of the voltage source.

$$R_t = \frac{R1 \times R2}{R1 + R2}$$

5. **TWO PARALLEL RESISTORS** — The formula for finding the effective resistance of a combination of two parallel resistors is

$$R_t = \frac{R1 \times R2}{R1 + R2}$$

$$\frac{1}{R_t} = \frac{1}{R1} + \frac{1}{R2} + \frac{1}{R3}$$

6. **THREE OR MORE PARALLEL RESISTORS** — The formula for finding the effective resistance of a combination of three or more resistors connected in parallel is

$$\frac{1}{R_t} = \frac{1}{R1} + \frac{1}{R2} + \frac{1}{R3} \cdots$$

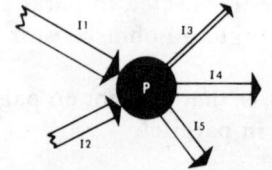

7. **KIRCHHOFF'S FIRST LAW** — The sum of all the currents flowing toward a junction always equals the sum of all the currents flowing away from that junction.

Self-Test— Review Questions

1. Calculate the equivalent value of resistance in each of the following parallel connections:
   (a) Two resistors of 12 and 8 ohms, respectively.
   (b) Six resistors, each valued at 4.8 ohms.
   (c) Two resistors of 20 and 4 K, respectively.
   (d) Two resistors of 1 M and 1.5 M, respectively.
2. Calculate the equivalent value of resistance for each of the following parallel connections:
   (a) Three resistors of 20, 30, and 40 ohms, respectively.
   (b) Four resistors of 20, 30, 40, and 50 ohms, respectively.
   (c) Three resistors of 20, 30, and 40 K, respectively.
3. Sketch the *equivalent circuit* of the parallel connection shown below:

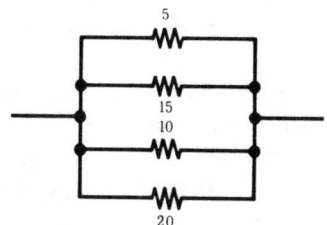

4. What value of resistance would you need to connect in parallel with a 20-ohm resistor to produce an equivalent resistance of 12 ohms?
5. A 22-K resistor having a tolerance of 10% has failed. You have on hand five resistors, valued at 120, 220, 47, 33, and 120 K, respectively. What combination of these would you choose to connect in parallel to replace the failed component? Give your answer by sketching the circuit diagram of the resultant.
6. In the circuit shown, what is value and direction of I3?

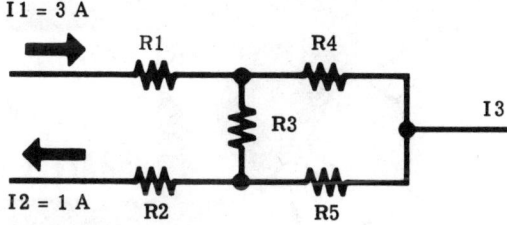

7. Sketch the equivalent circuit of

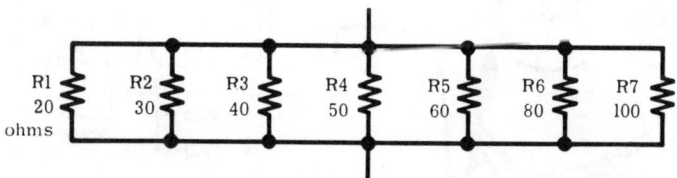

Applying Ohm's Law in Parallel Circuits

You have already seen something of the use of Ohm's Law in a parallel circuit. In practice, a good many more unknown quantities of current, voltage, and resistance in such circuits can generally be calculated by using this law.

Suppose you wanted to find out the *resistance* of a resistor connected in parallel with one or more other resistors. If you used an ohmmeter, you would first have to disconnect the resistor to be measured from the circuit; otherwise the ohmmeter would read the total resistance of the parallel combination of resistors.

Again, if you were to set about measuring with an ammeter the *current flow* through one particular resistor of a combination of parallel resistors, you would first have to disconnect it, and insert the ammeter to read only the current flow through that particular resistor.

In either case, time and effort could often be saved by the use of Ohm's Law.

If you were trying to find the *voltage* existing across a parallel circuit, of course, a direct voltmeter reading could be obtained without any need for a disconnection; but here again intelligent application of Ohm's Law will often give you the information you want without the need for actual measurement.

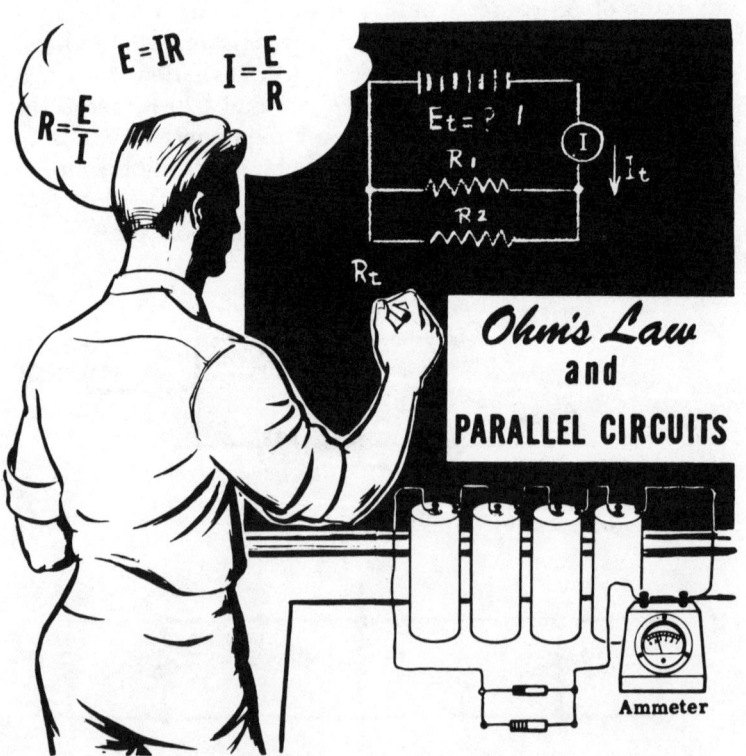

Solving Unknowns in Parallel Circuits

Six facts, all of them known to you by now, are all the equipment you need for finding out the unknown values in a dc parallel circuit.

In the circuit opposite, a voltage E is applied across three resistors —R1, R2, and R3—connected in parallel. What are the facts you know about this circuit?

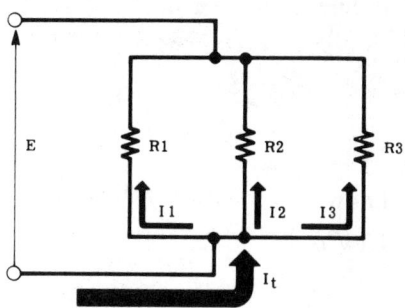

FACT 1: Kirchhoff's First Law tells you that

$$I_t = I_1 + I_2 + I_3$$

FACT 2: You know that the full circuit voltage (E) appears across each one of the three parallel resistors.

FACT 3: Fact 2, plus Ohm's Law, tells you that

$$I_1 = \frac{E}{R1} \quad I_2 = \frac{E}{R2} \quad I_3 = \frac{E}{R3}$$

FACT 4: Any parallel circuit can be reduced to an equivalent circuit.

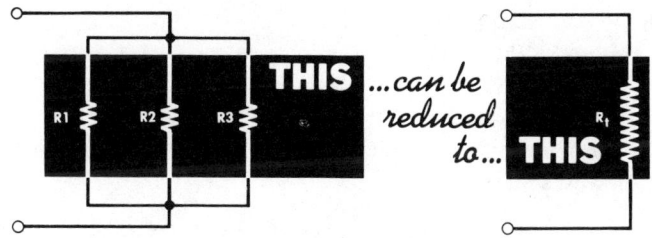

FACT 5: Ohm's Law can then be applied to the equivalent circuit.

$$I_t = \frac{E}{R_t} \quad \text{or} \quad R_t = \frac{E}{I_t} \quad \text{or} \quad E = I_t R_t$$

FACT 6: The total resistance of any parallel circuit can be found by applying the formula

$$\frac{1}{R_t} = \frac{1}{R1} + \frac{1}{R2} + \frac{1}{R3} + \ldots$$

and for 2 resistors in parallel

$$R_t = \frac{R1 \times R2}{R1 + R2}$$

Solving Unknowns in Parallel Circuits (continued)

Now watch how these six facts can be used to solve the sort of problem which you will be constantly meeting in your practical work in electricity and electronics.

Problem 1

Two resistors in a circuit are connected in parallel. One of them is marked with the value *15 ohms*, but the other is unmarked in any way. You lack an ohmmeter, but have both a voltmeter and an ammeter. With these instruments, you read that the voltage across the parallel combination is 6 volts and that the current flowing into the combination is 1 ampere. You need to know the value of the unmarked resistor.

Solution. First, sketch the circuit, and fill in the values you know.

You see at once that you know two of the three Ohm's Law quantities, and that $R_t = \dfrac{E}{I_t} = \dfrac{6}{1}$. The total resistance of the parallel circuit is, therefore, 6 ohms.

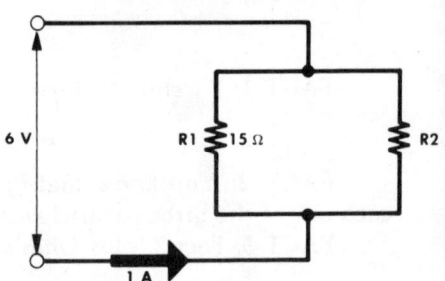

You already know the value of R1 (it is marked *15 ohms*). So take Fact 6 (the Parallel Resistance Formula), and substitute known values wherever you can.

$$R_t = \frac{R1 \times R2}{R1 + R2}$$

Therefore, in this case,

$$6 = \frac{15 \times R2}{15 + R2}$$

Now multiply both sides of this last equation by the factor (15 + R2), and you get

$$6 \times (15 + R2) = 15 \times R2$$

Multiply out the left-hand side of the equation:

$$90 + 6R2 = 15R2$$

Work this equation out according to the simple laws of algebra, and you see that

$$15R2 - 6R2 = 90$$

Therefore, $9R2 = 90$

and $R2 = 10$ ohms

You know now that the value of the unmarked resistor must be 10 ohms—and you have solved a typical problem of the kind you will often meet in practice by selecting one or more of the six facts which you needed, and by applying to them a little simple mathematics.

Solving Unknowns in Parallel Circuits (continued)

Try another practical problem of the same kind.

Problem 2

You are faced with a circuit having three resistors with values of 3, 9, and 12 ohms, respectively, connected in parallel. You find it possible to measure with an ammeter that the current flowing through the 9-ohm resistor is 8 amperes, but the other two resistors are inaccessible. You need to know the circuit current (possibly because you want to connect into the circuit a load which will burn out if too much current flows through it, and you need to find out the value of the resistor you will have to insert into the circuit to protect this load).

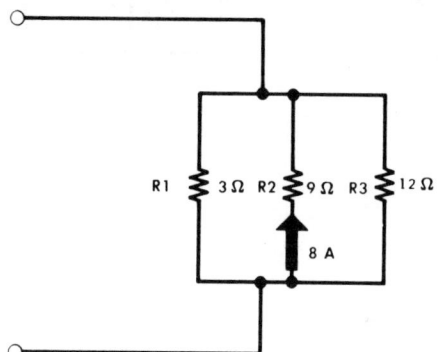

Sketch the circuit, and fill in on it the values you know. You have two of the essential facts about R2, so use Ohm's Law to calculate the third:

$E = I_2 \times R_2$
$= 9 \text{ ohms} \times 8 \text{ amperes}$
$= 72 \text{ volts}$

The voltage across one branch of a parallel circuit is the voltage across all of them; so you now have the facts you need to know about R1 and R3 to use Ohm's Law to find the current flowing through each of them.

Take R1 first. You know that its value is 3 ohms, and that the voltage across it is 72 volts. So,

$$I_1 = \frac{E}{R_1} = \frac{72}{3} = 24 \text{ amperes}$$

Then deal with R3. Its value is 12 ohms, and the voltage through it is still 72 volts. So,

$$I_3 = \frac{E}{R_3} = \frac{72}{12} = 6 \text{ amperes}$$

You now know the current flowing through all three branches of the parallel circuit. Kirchhoff's Second Law tells you that the circuit current itself is the sum of these three currents ($I_t = I_1 + I_2 + I_3$). So, in this case,

$$I_t = 8 + 24 + 6 = 38 \text{ amperes}$$

which is the value of the circuit current you are looking for.

Review of Ohm's Law and Parallel Circuits

If a circuit consists of two or more resistors—R1, R2, R3, etc.—in parallel, the following rules for using Ohm's Law apply:

R_t, I_t, and E are used together.
R1, I1, and E are used together.
R2, I2, and E are used together; and so on.

Only quantities having the same, or no, subscript can be used together to find an unknown by means of Ohm's Law.

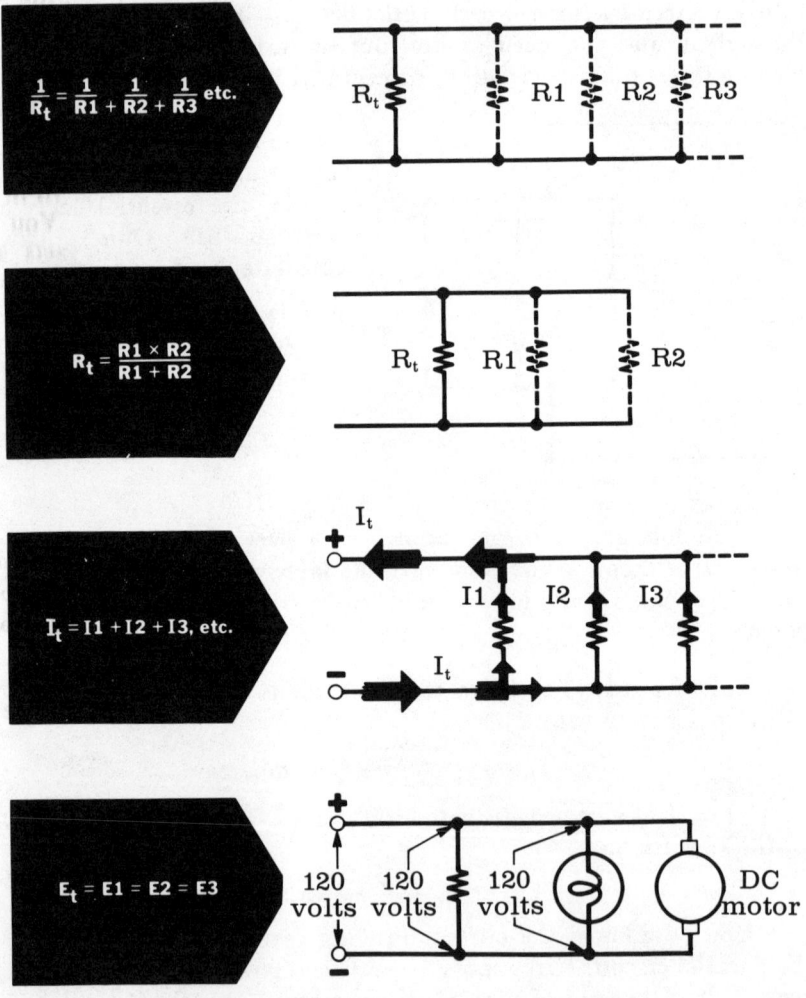

and since E = IR
$I_1 \times R1 = I_2 \times R2 = I_3 \times R3$ etc.

DC PARALLEL CIRCUITS

Self-Test—Review Questions

1. Three resistors—R1 = 16 ohms, R2 = 24 ohms, and R3 = 32 ohms—are connected in parallel; the total current passing through them is 5.2 amperes.
   (a) What is the current passing through each resistor?
   (b) What is the voltage across the parallel combination?
2. Four resistor loads are connected in parallel across a 120-volt line. Their values are 500, 200, 100, and 50 ohms, respectively. What is the total current drawn? What are the individual currents?
3. A divider chain has been built to work into a known load as follows:

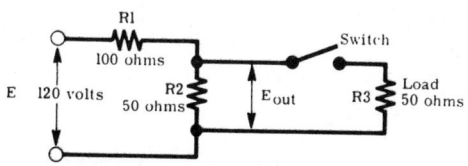

   (a) What is E_{out} with the switch open?
   (b) What is E_{out} with the switch closed?
4. Solve the circuits shown below for the unknowns:

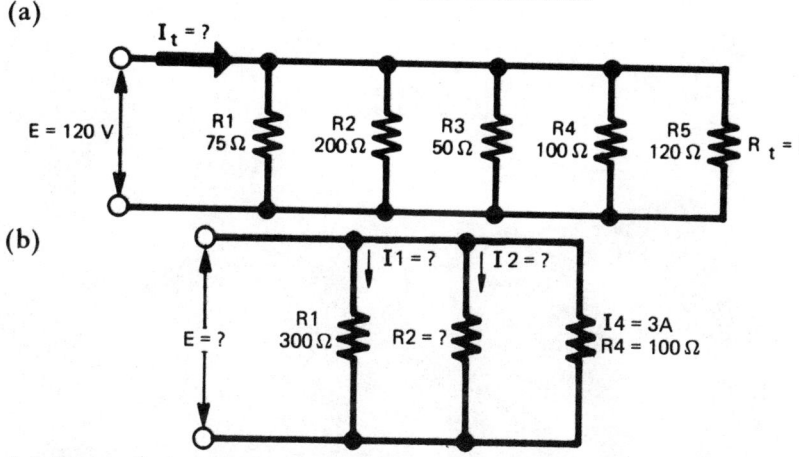

   (c) Draw the equivalent circuits for (a) and (b) above.
5. A moving-coil meter movement has a resistance of 10 ohms and needs a current of 40 mA to give it full-scale deflection. How would you adapt such an instrument in order to use it as an ammeter reading up to 0.2 ampere?
6. A moving-coil meter movement requires a current of 25 mA to give it full-scale deflection. The voltage required to produce this deflection is 25 mV. How would you adapt this meter for use as
   (a) a millivoltmeter capable of reading 0-100 mV?
   (b) a voltmeter capable of reading 0-200 V?
   (c) a milliammeter capable of reading 0-50 mA?
   (d) an ammeter capable of reading 0-50 A?

Experiment/Application—Ohm's Law and Parallel Resistances

To see how an ammeter and voltmeter may be used as a substitute for an ohmmeter to find the values of the individual and total resistances in a parallel combination, four dry cells can be connected in series to be used as a voltage source. Then connect a voltmeter across the dry cell battery to make certain the voltage remains constant at 6 volts, and connect an ammeter in series with the negative (−) terminal of the battery to read the current.

Now if you connect a fuse, a resistor with one orange band and two black ones, and a switch in series between the positive (+) terminals of the ammeter and the battery, you will see that the voltage remains at 6 volts, and that the current indicated is 0.2 ampere. From Ohm's Law, the resistance value must be 30 ohms, and a check of the color code shows that this is, indeed, the correct value.

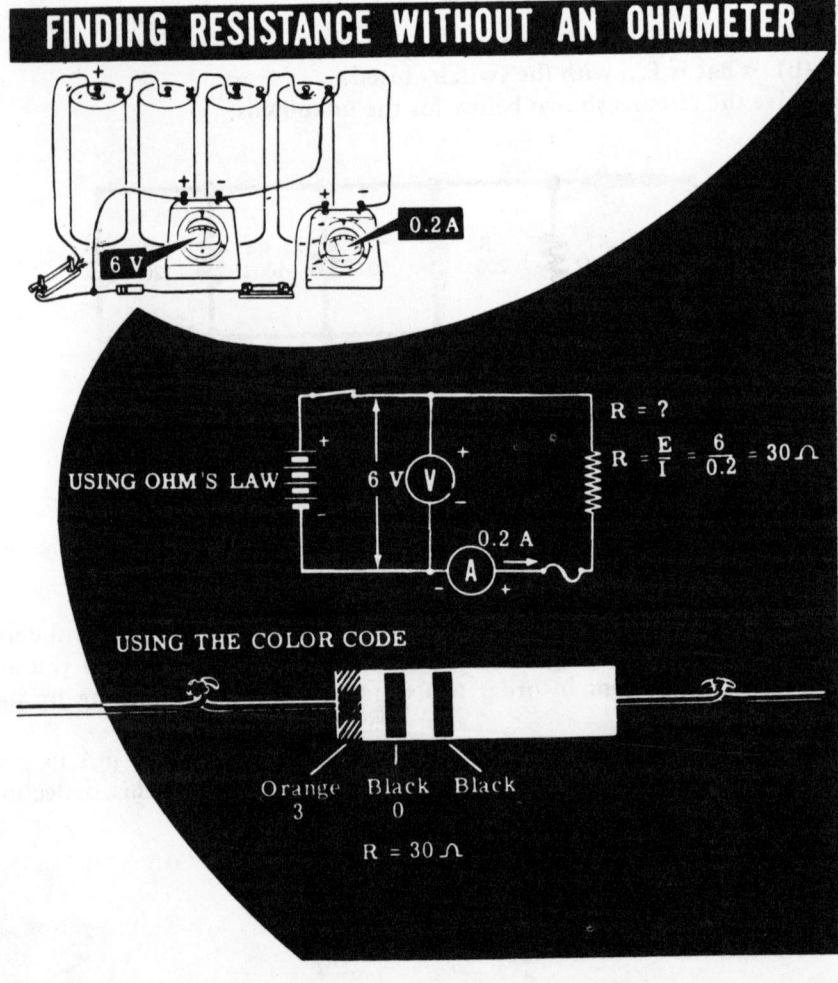

Experiment/Application—Ohm's Law and Parallel Resistances (continued)

If another resistor carrying one brown, one green, and one black band is added in parallel, you see that the current reading is 0.6 ampere, with no change in voltage. Since the first resistor passes 0.2 ampere, the current through the second resistor must be 0.4 ampere. So the Ohm's Law value of the second resistor is 6 volts divided by 0.4 ampere, or 15 ohms.

The total resistance of the parallel combination is equal to 6 volts divided by the total current, 0.6 ampere—or 10 ohms.

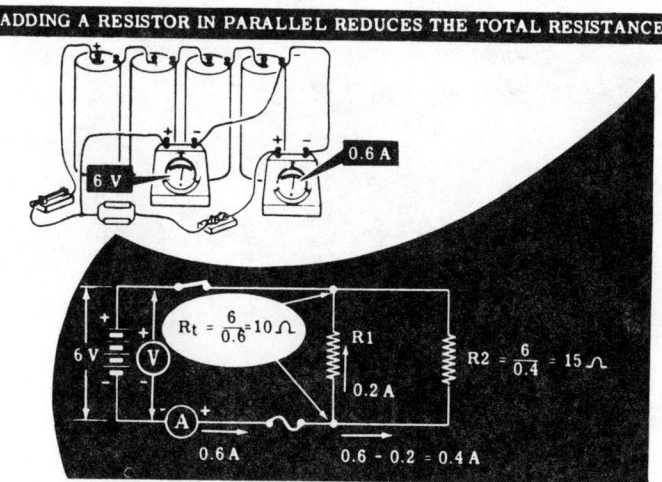

With still another resistor (one orange and two black bands) added in parallel, you see a further 0.2-ampere increase in current—showing that the Ohm's Law value of the added resistor is 30 ohms. The total current is now 0.8 ampere, resulting in a total resistance value of 7.5 ohms for the parallel combination.

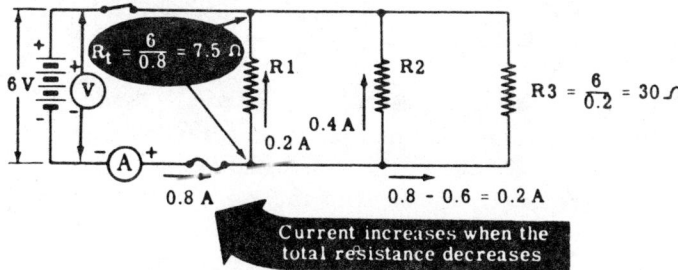

When you have disconnected the battery and the various resistors, you can check the total and the individual resistances with an ohmmeter; you will see that the Ohm's Law and color code values are identical with the values indicated by the ohmmeter.

DC PARALLEL CIRCUITS

Experiment/Application—Ohm's Law and Parallel Circuit Current

Using only three series-connected cells as a voltage source, you could connect the voltmeter across the battery. Then you could connect four resistors—two 15-ohm and two 30-ohm resistors—across the battery in parallel.

By Ohm's Law, the current through each 15-ohm resistor will be 0.3 ampere, and through each 30-ohm resistor, 0.15 ampere. The total current will be the sum of the currents through the individual resistances, or 0.9 ampere.

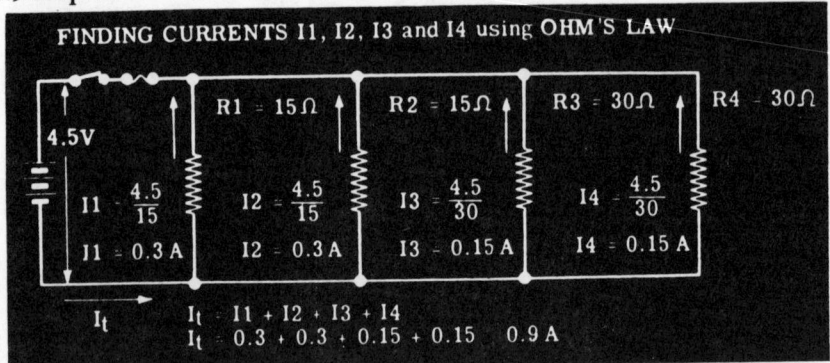

MEASURING CURRENTS IN PARALLEL CIRCUITS

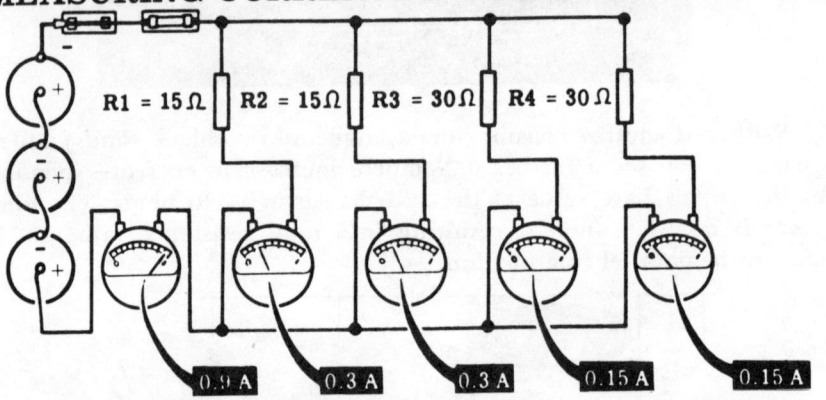

If you now insert an ammeter in the circuit—first to read the total circuit current, then that of the individual resistances—you will see that the actual currents are the same as those found by applying Ohm's Law, and that total current equals the sum of all the individual currents.

Learning Objectives—Next Section

Overview—You are now ready to proceed with solving complex series-parallel circuits. You will see that any complex circuit can be reduced to solvable combinations of series and parallel circuits.

Series-Parallel Circuits

Circuits consisting of three or more resistors may be connected in a complex circuit, partially series and partially parallel.

There are two basic types of series-parallel circuits: one in which a resistance is connected in series with a parallel combination; and the other in which one or more branches of a parallel circuit consist of resistances in series.

If you were to connect two lamps in parallel (side-by-side connection) and connect one terminal of a third lamp to one terminal of the parallel combination, the three lamps would be connected in series-parallel. Resistances other than lamps may also be connected in the same manner to form series-parallel circuits.

You can connect the three lamps to form another type of series-parallel circuit by first connecting two lamps in series, then connecting the two terminals of the third lamp across the series lamps. This forms a parallel combination with one branch of the parallel circuit consisting of two lamps in series.

Such combinations of resistance are frequently used in electric circuits, particularly in electric motor circuits and in control circuits for electrical equipment.

TWO WAYS OF CONNECTING LAMPS IN SERIES-PARALLEL

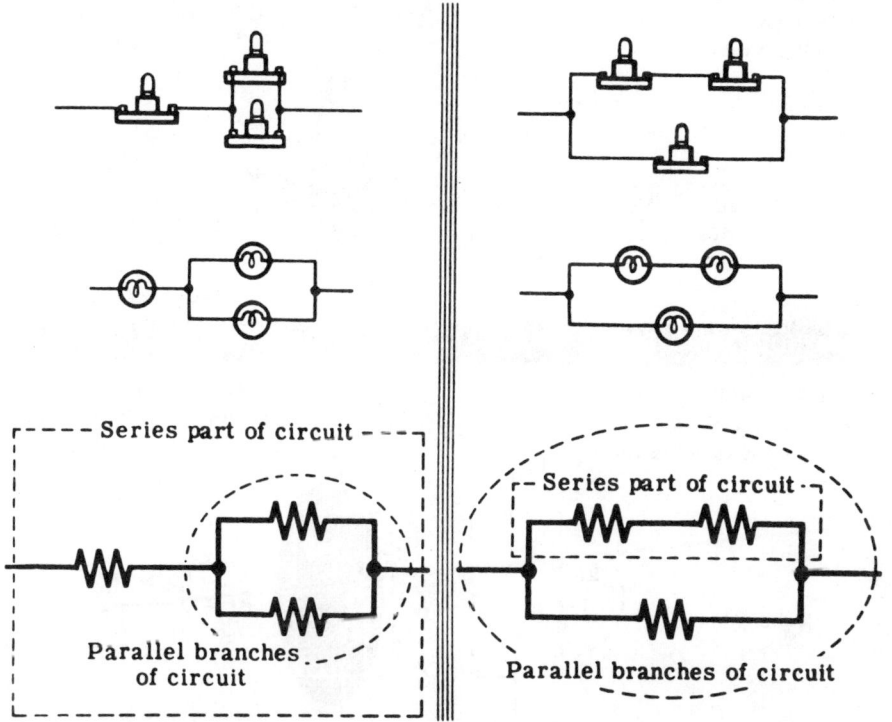

Resistors in Series-Parallel

No new formulas are needed to find the total resistance of resistors connected in series-parallel. What you need to do is to break the complete circuit into parts, each consisting of simple series and simple parallel circuits. Then solve each part separately and combine the answers. But before using the rules for series and parallel resistances, you must first decide how best to simplify the circuit.

Suppose your problem is to find the total resistance of three resistors—R1, R2, R3—connected in series-parallel, with R1 and R2 connected in parallel, and R3 connected in series with the parallel combination. To simplify the circuit, you would break it down into two parts—the parallel circuit of R1 and R2, and the series resistance R3. First you find the equivalent resistance of R1 and R2, using the formula for parallel resistances. This value is then added to the series resistance R3 to find the total resistance of the series-parallel circuit.

If the series-parallel circuit consists of R1 and R2 in series, with R3 connected across them, the steps are reversed. The circuit is broken down into two parts—the series circuit of R1 and R2, and the parallel resistance R3. First you find the total resistance of R1 and R2 by adding; then combine this value with R3, using the formula for parallel resistance.

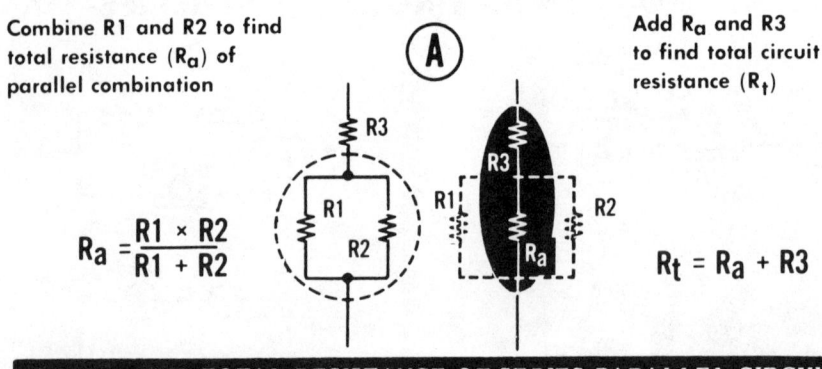

Combine R1 and R2 to find total resistance (R_a) of parallel combination

$$R_a = \frac{R1 \times R2}{R1 + R2}$$

Add R_a and R3 to find total circuit resistance (R_t)

$$R_t = R_a + R3$$

FINDING THE TOTAL RESISTANCE OF SERIES-PARALLEL CIRCUIT

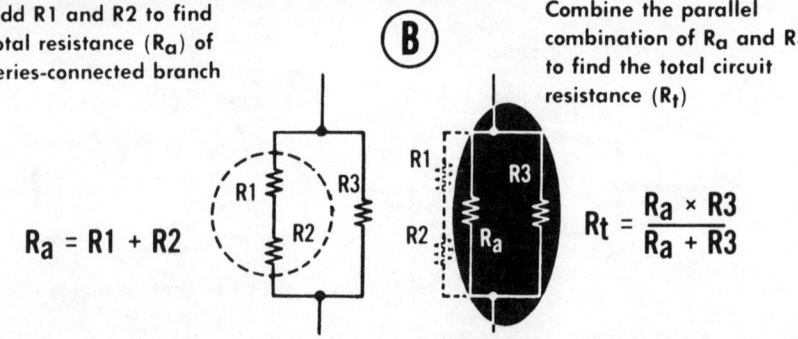

Add R1 and R2 to find total resistance (R_a) of series-connected branch

$$R_a = R1 + R2$$

Combine the parallel combination of R_a and R3 to find the total circuit resistance (R_t)

$$R_t = \frac{R_a \times R3}{R_a + R3}$$

Resistors in Series-Parallel (continued)

Complex circuits may be simplified and their breakdown made easier by redrawing the circuits before applying the steps to combine resistances.

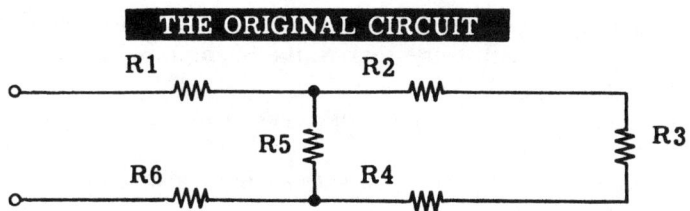

1. Start at one end of the circuit and draw all series resistances in a vertical straight line until you reach a point where the circuit has more than one path to follow. At that point draw a horizontal line across the end of the series resistance.

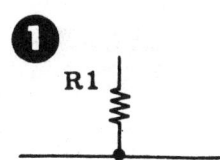

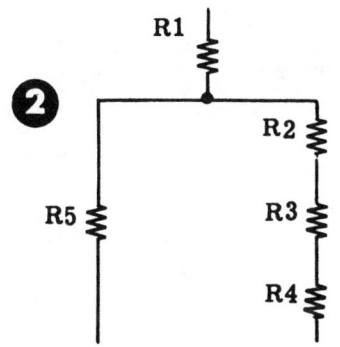

2. Draw the parallel paths from this line in the same direction as the series resistances—that is, vertically.

3. Where the parallel paths combine, draw another horizontal line across the ends to join the paths.

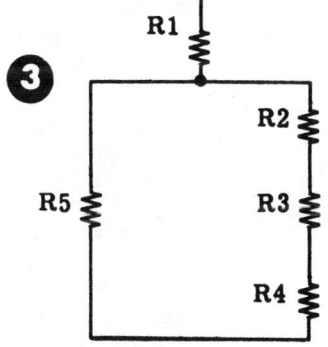

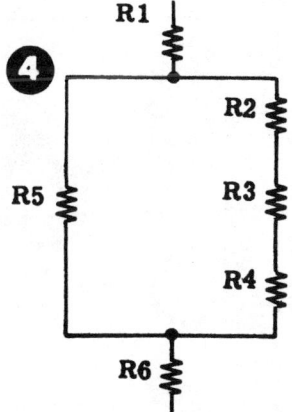

4. Continue the circuit from the center of the parallel connecting line, adding the series resistance to complete the redrawn circuit.

Resistors in Series-Parallel (continued)

The basic steps in finding the total resistance of a complex series-parallel circuit are therefore as follows:

1. Redraw the circuit if necessary.
2. If any of the parallel combinations have branches consisting of two or more resistors in series, find the total value of these resistors by adding them.
3. Using the formula for parallel resistances, find the total resistance of the parallel parts of the circuits.
4. Add the combined parallel resistances to any resistances which are in series with them.

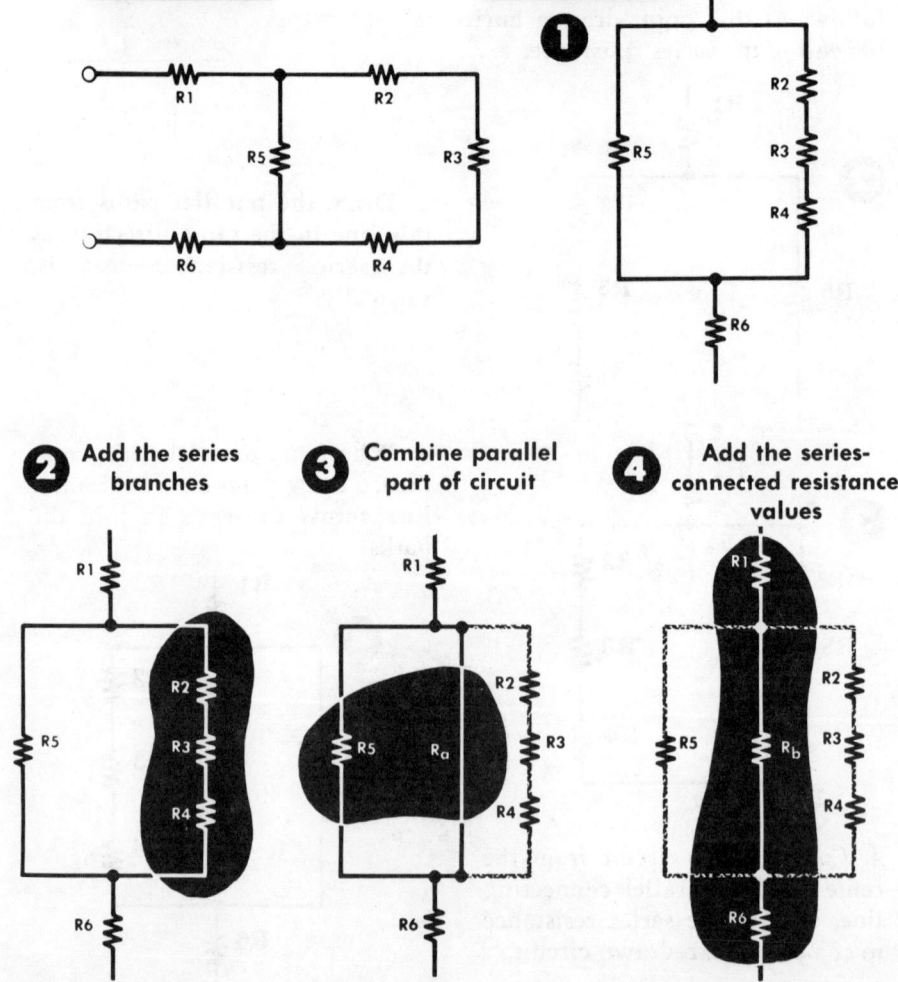

DC SERIES-PARALLEL CIRCUITS

Resistors in Series-Parallel (continued)

Here is a practical example of how to break down complex circuits to find the total resistance:

Suppose your circuit consists of four resistors—R1, R2, R3, and R4—connected as shown. You want to find the total resistance of the circuit.

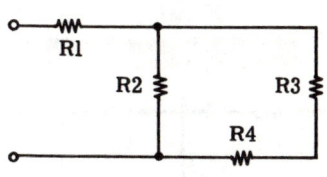

Suppose also that R1 = 7 ohms, R2 = 10 ohms, R3 = 6 ohms, and R4 = 4 ohms.

First, the circuit is redrawn and the series branch resistors R3 and R4 are combined by addition to form an equivalent resistance R_a.

R_a = R3 + R4
    = 6 + 4
    = 10 ohms

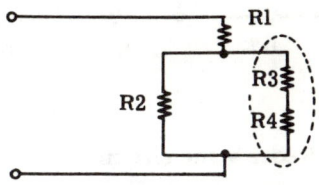

Next, the parallel combination of R2 and R_a is combined (using the parallel resistance formula) as an equivalent resistance, R_b.

$$R_b = \frac{R2 \times R_a = 10 \times 10}{R2 + R_a = 10 + 10}$$
    = 5 ohms

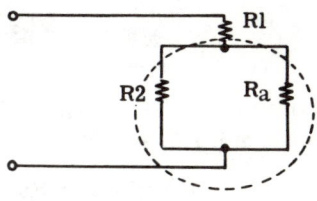

Last, the series resistor R1 is added to the equivalent resistance —R_b—of the parallel combination to find the total circuit resistance, R_t.

R_t = R1 + R_b
    = 7 + 5
    = 12 ohms

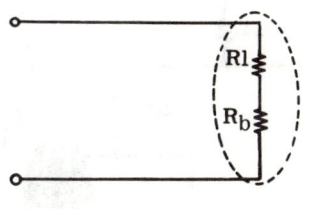

In other words, the whole complex circuit can be broken down and simplified until R_t = total resistance of series-parallel circuit = 12 ohms.

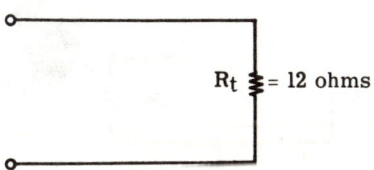

Resistors in Series-Parallel (continued)

More complicated circuits only require more steps; they do not require any additional formulas. For example, the total resistance of a circuit consisting of nine resistors may be found as shown:

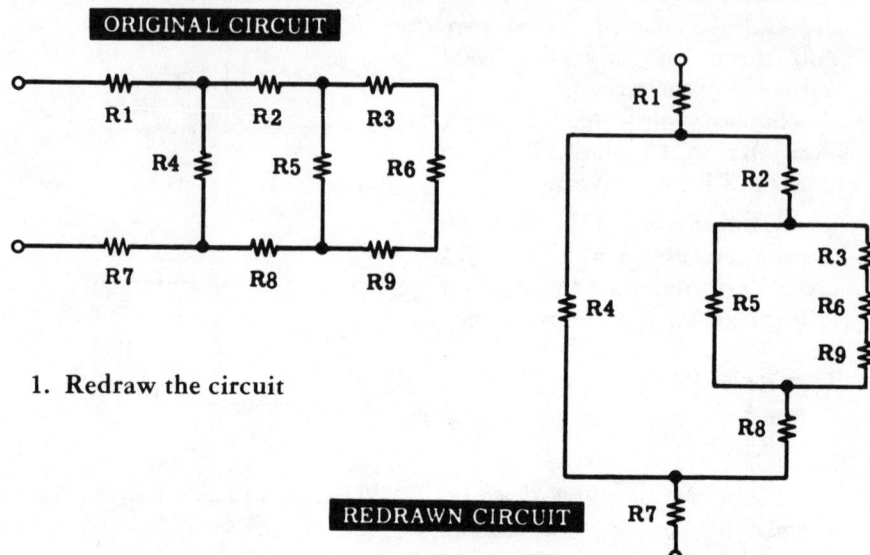

1. Redraw the circuit

2. Combine the series branch resistors R3, R6, and R9.

$$R_a = R3 + R6 + R9$$

3. Combine the parallel resistances R5 and R_a.

$$R_b = \frac{R5 \times R_a}{R5 + R_a}$$

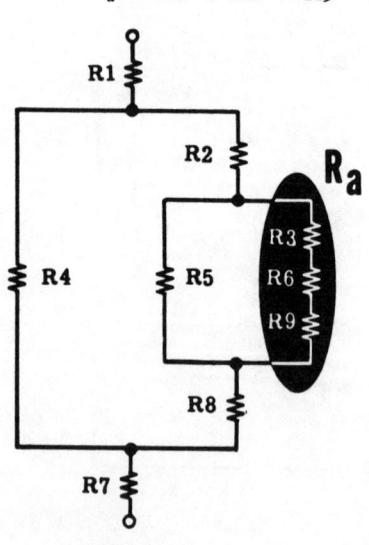

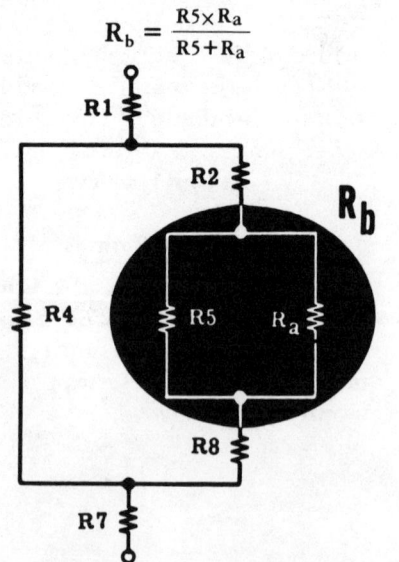

Resistors in Series-Parallel (continued)

4. Combine the series resistances R2, R_b, and R8.

$$R_c = R2 + R_b + R8$$

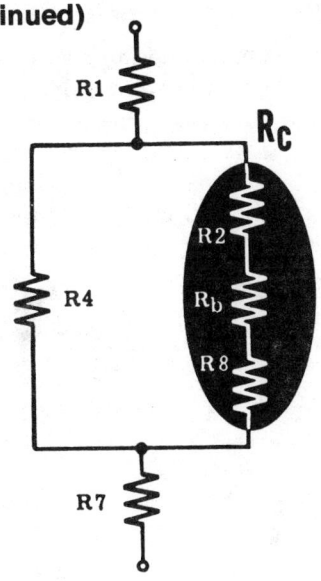

5. Combine the parallel resistances R4 and R_c.

$$R_d = \frac{R4 \times R_c}{R4 + R_c}$$

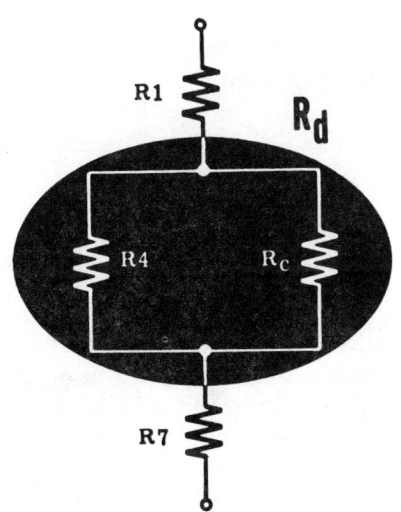

6. Combine the series resistances R1, R_d, and R7.

$$R_t = R1 + R_d + R7$$

7. R_t is the total resistance of the circuit, and the circuit will act as a single resistor of this value when connected across a voltage source.

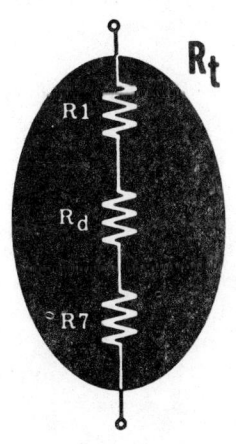

Solving the Bridge Resistor Circuit

There remains one important type of complex circuit which you do not yet know how to solve easily.

Look at the circuit below. Its outline is familiar enough—but you see that there is an extra resistor (R2) connecting the two parallel branches of the series-parallel combination in such a way that the series connection in both branches is interrupted by the leads to the new resistor. This new resistor—R2—is known as a *bridge*.

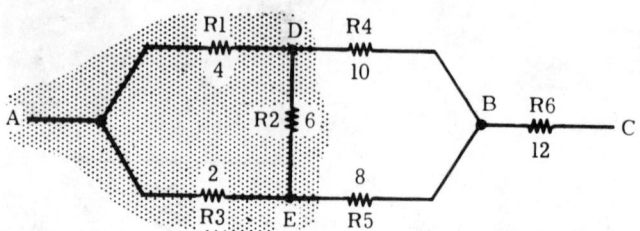

If you look at the shaded part of the circuit above you see that it is essentially the shape you see opposite. This arrangement, from its similarity to the shape of the Greek letter D (delta), is said to be *delta-connected*.

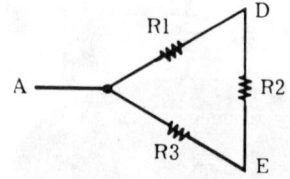

You also observe, however, that if you could devise a circuit shaped like a Y (wye) *such that its terminal resistances at D and E were identical in value* with the corresponding terminal resistances in the delta circuit, this new Y circuit would fit onto the rest of the original circuit in such a way that you could solve its values without difficulty. Look at the diagram below.

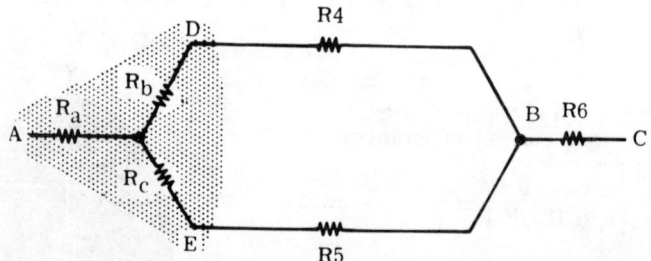

Call the three resistors in your proposed Y circuit R_a, R_b, and R_c. Remember that their values must be such that the terminal resistances at D and E are exactly what they were in the original circuit.

Your problem is to find a formula for expressing R_a, R_b, and R_c, whose values you don't know, in terms of R1, R2, and R3, whose values you do know.

Solving the Bridge Resistor Circuit (continued)

Redraw the delta circuit and the Y circuit from the last page together so that you can look at them conveniently side-by-side. Then mark in firmly the *equals* sign between them to remind you that both circuits must give exactly the same values of resistance across every corresponding pair of terminals. You are now all set for an operation called the *delta-Y conversion*.

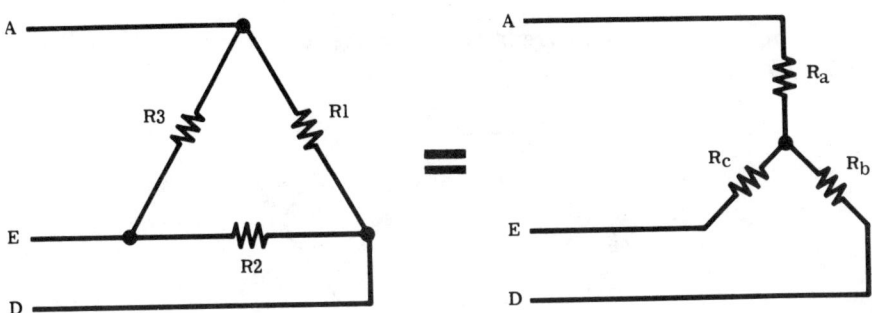

Consider, first, the sum of the resistances between A and E, assuming D to be disconnected. In the delta combination you will see that between these two points there is effectively a series combination of R1 and R2 in parallel across R3. You can, therefore, from the knowledge you already have, express the resistance A-E as

$$\frac{R3 \ (R1 + R2)}{R1 + R2 + R3}$$

In the Y circuit, the total resistance between A and E is obviously $R_a + R_c$. Since you know that these two resistances must be equal, you can write down as your first equation:

$$R_a + R_c = \frac{R3 \ (R1+R2)}{R1+R2+R3}$$

In exactly the same way, you can express the total resistances between A-D and between D-E in terms of R1−R2−R3 and of R_a−R_b−R_c. Work it out for yourself, and you will get two more equations as follows:

$$R_a + R_b = \frac{R1 \ (R2+R3)}{R1+R2+R3}$$

$$R_c + R_b = \frac{R3 \ (R1+R2)}{R1+R2+R3}$$

Now do a little simple algebra (beginning by subtracting equation (2) from equation (1), to get equation (4); then adding equation (4) to equation (3), to get a value for R_c in terms of R1 - R2 - R3; and, lastly, substituting for R_c in equations (1) and (3) to get similar values for R_a and R_b). You find that:

$$\boxed{R_a = \frac{R1 \times R3}{R1+R2+R3} \ ; \ R_b = \frac{R1 \times R2}{R1+R2+R3} \ ; \ R_c = \frac{R2 \times R3}{R1+R2+R3}}$$

Solving the Bridge Resistor Circuit (continued)

Now go back to the original circuit, and fill in the known values of R1, R2, and R3. You get:

$$R_a = \frac{4 \times 2}{4+6+2} = 0.67 \text{ ohm}; \quad R_b = \frac{4 \times 6}{4+6+2} = 2 \text{ ohms};$$

$$R_c = \frac{2 \times 6}{4+6+2} = 1 \text{ ohm}$$

The original circuit (redrawn, with the Y connection shaded) now looks like this:

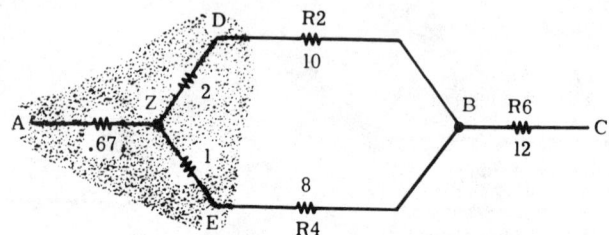

Here you have, between Z and B, a simple parallel circuit whose equivalent resistance you already know how to calculate.

$$R_{zb} = \frac{(2 + 10) \times (1 + 8)}{2 + 10 + 1 + 8} = \frac{108}{21} = 5.14 \text{ ohms}$$

The equivalent resistance of the entire bridge circuit is therefore:

$$AZ + ZB + BC = 0.67 + 5.14 + 12 = 17.81 \text{ ohms}$$

correct to two decimal places. Your problem is solved.

It is quite easy to convert any delta connection into its Y equivalent if you get the numbering of the various resistors right. Look at the drawing below, in which the equivalent Y is dotted in on top of an original delta:

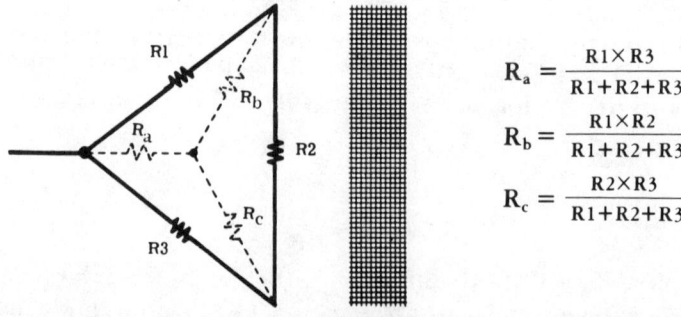

Label as R_a the Y connection which bisects the angle R1-R3; label as R_b the connection which bisects R1-R2; and label as R_c the connection which bisects R2-R3. If you do this, the formula almost remembers itself. The denominator is always R1 + R2 + R3. The numerator is the product of the two delta resistors which your Y resistor connection bisects.

Ohm's Law in Series-Parallel Circuits—Current

The total circuit current for a series-parallel circuit depends on the total resistance offered by the circuit when connected across a voltage source. Current flow in the circuit will divide to flow through all parallel paths, and come together again to flow through series parts of the circuit. It will divide to flow through a branch circuit, and then repeat this division if the branch circuit itself subdivides into secondary branches.

As in parallel circuits, the current through any branch resistance is inversely proportional to the resistance of the branch—the greater current flows through the least resistance. However, all of the branch currents always add up to equal the total circuit current.

The total circuit current is the same at each end of a series-parallel circuit, and is equal to the current flow through the voltage source.

HOW CURRENT FLOWS IN A SERIES-PARALLEL CIRCUIT

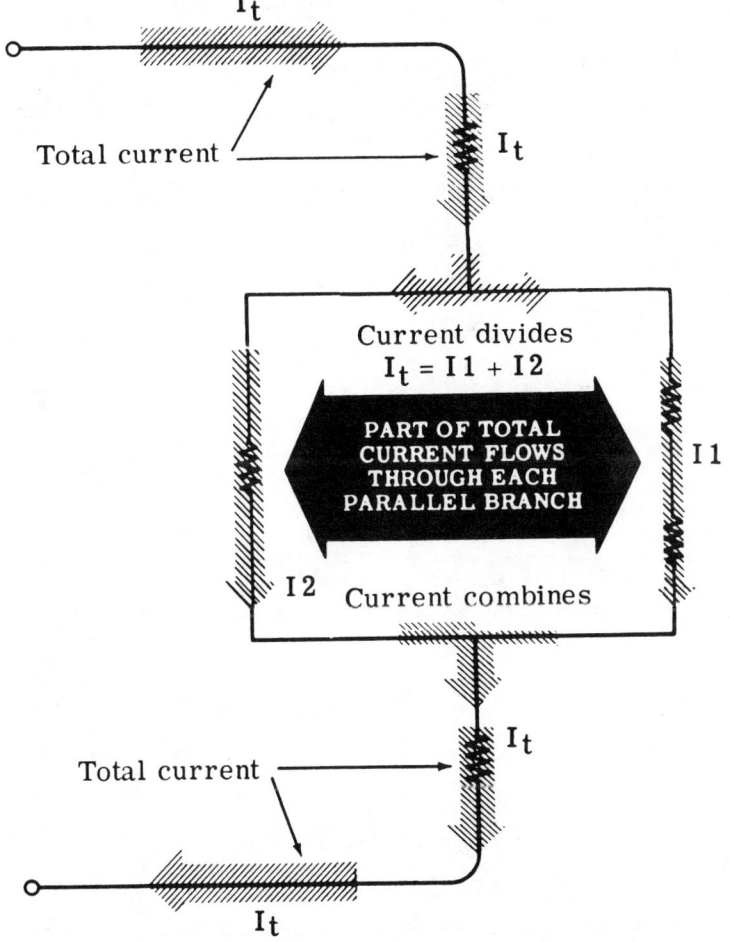

Ohm's Law in Series-Parallel Circuits—Voltage

Voltage drops across a series-parallel circuit occur in the same way as they do in series and parallel circuits. In the series parts of the circuit, the voltage drops across the resistors depend on the individual values of the resistors. In the parallel parts of the circuit, every branch has the same voltage across it, and carries a current which is dependent on the resistance in that particular branch.

Series resistances forming a branch of a parallel circuit will divide the voltage across the parallel circuit. In a parallel circuit consisting of a branch with a single resistance and a branch with two series resistances, the voltage across the single resistance is equal to the sum of the voltages across the two series resistances. The voltage across the entire parallel circuit is exactly the same as that across either of the branches.

The voltage drops across the various paths between the two ends of the series-parallel circuit always add up to the total voltage applied to the circuit.

HOW THE VOLTAGE DIVIDES IN A SERIES-PARALLEL CIRCUIT

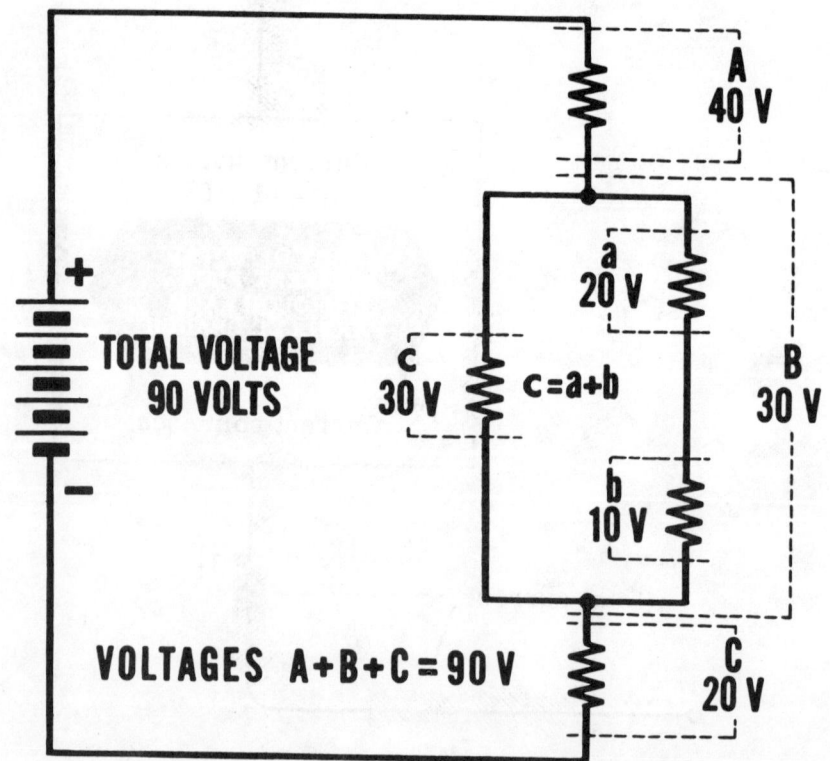

Ohm's Law in Series-Parallel Circuits

Example 1

Consider the circuit of page 2-92:

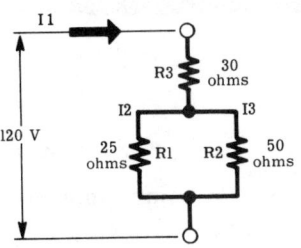

To find the current I_1 through this circuit

$$R_a = \frac{R1 \times R2}{R1 + R2} = \frac{25 \times 50}{25 + 50} = \frac{1,250}{75} = 16.67 \text{ ohms}$$

$$R_t = 46.67 \text{ ohms}$$

$$I_1 = \frac{120}{46.67} = 2.57 \text{ amperes}$$

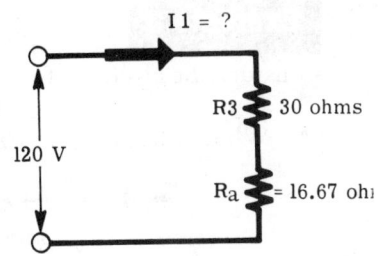

Example 2

Consider the circuit of pages 2-93 and 2-94:

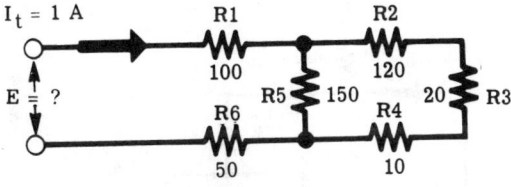

Redrawn, the circuit looks like

Ohm's Law in Series-Parallel Circuit (continued)

Example 2 (cont.)

$R_a = R2 + R3 + R4 = 120 + 20 + 10 = 150$

$R_b = \dfrac{R_a \times R5}{R_a + R5} = \dfrac{150 \times 150}{150 + 150} = 75$

$R_t = R1 + R_b + R6 = 100 + 75 + 50 = 225$

Using Ohm's Law, the voltage can be calculated as

$E = I_t \times R_t = 1A \times 225 = 225$ volts

Example 3

Consider the circuits of pages 2-96 and 2-97:

(A) What is E_{out} if E_{in} is 100 volts?

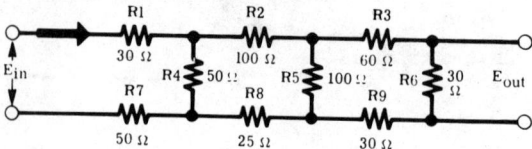

(B) Redrawn circuit (see page 2-97).

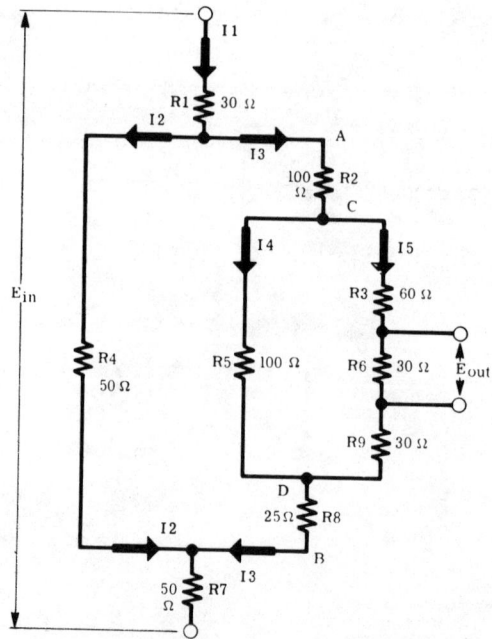

Ohm's Law in Series-Parallel Circuits (continued)

Example 3 (cont.)

To solve this, we must find the current through R6 to find E_{out}. Initially, we must find the currents which require us to find the resistance of the various paths. For the series combination of R3, R6, and R9,

$$R_a = R3 + R6 + R9 = 120 \text{ ohms}$$

Then R5 in parallel with R_a can be calculated as

$$R_b = \frac{R5 \times R_a}{R5 + R_a} = \frac{100 \times 120}{100 + 120} = \frac{12,000}{220} = 54.5 \text{ ohms}$$

The circuit can now be drawn as

(C)

Combining the series resistors R2, R_b, and R8 yields

$$R_{ab} = R_c = R2 + R_b + R8 = 100 + 54.5 + 25 = 179.5$$

The parallel combination of R4 and R_c is

$$R_d = \frac{R_c \times R4}{R_c + R4} = \frac{179.5 \times 50}{179.5 + 50}$$
$$= \frac{897.5}{229.5} = 39.1$$

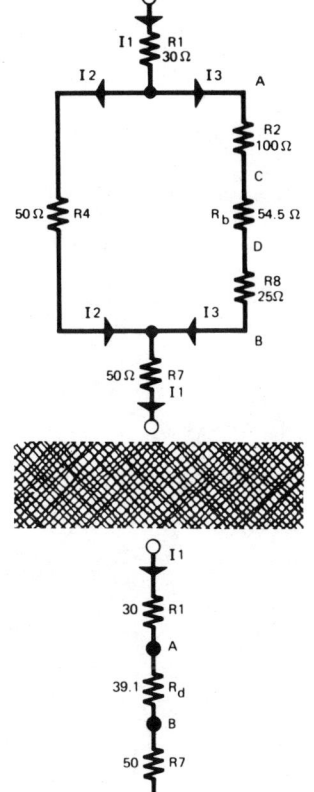

The circuit can now be drawn as (D)

And the current is calculated as

$$I_1 = \frac{E_{in}}{R_t}, \text{ where}$$
$$R_t = R1 + R_d + R7$$
$$= 30 + 39.1 + 50$$
$$= 119.1$$

$$I_1 = \frac{100}{119.1} = 0.84 \text{ ampere}$$

Ohm's Law in Series-Parallel Circuits (continued)

Example 3 (cont.)

We can now back track to drawing (C) of our circuit and calculate the currents I_2 and I_3.

The voltage drop across R_d—from point A to point B in drawing (C)—can be calculated as

$$E_{ab} = I_1 \times R_d = 0.84 \times 39.1 = 32.8 \text{ volts}$$

As shown in drawing (C), we can calculate the current I_3 as

$$I_3 = \frac{E_{ab}}{R_c} = \frac{32.8}{179.5} = 0.183 \text{ ampere}$$

But from drawing (B),

$$I_3 = I_4 + I_5$$

We can calculate E_{cd}, which we need to find the current I_5.

$$E_{cd} = I_3 \times R_b = 0.183 \times 54.5 = 9.97 \text{ volts}$$

Thus, $I_5 = \frac{E_{cd}}{R_a} = \frac{9.97}{120} = 0.83$ ampere

We can now finally solve for E_{out} since

$$E_{out} = I_5 \times R6 = 0.083 \text{ ampere} \times 30 = 2.49 \text{ volts}$$

While this may seem complicated, it is just the application of what we have been learning.

Measuring the Total Resistance of the Circuit...

Review of Series-Parallel Circuits

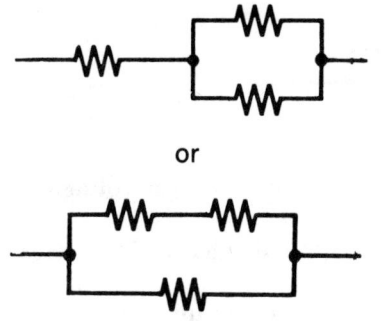

1. A SERIES-PARALLEL CIRCUIT —A series-parallel circuit has both parallel and series elements combined.

2. CIRCUIT REDUCTION—The formulas for determining series and parallel resistor combinations are used to reduce complex circuits.

3. CIRCUIT SIMPLIFICATION— Redrawing the circuit often results in simplification.

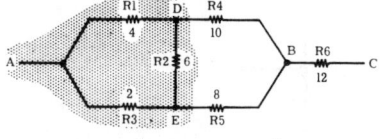

can be converted to

4. BRIDGE CONFIGURATION—

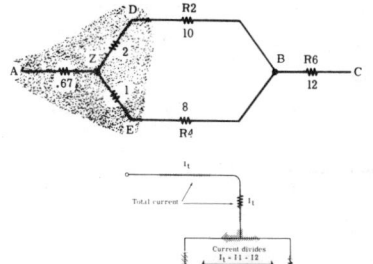

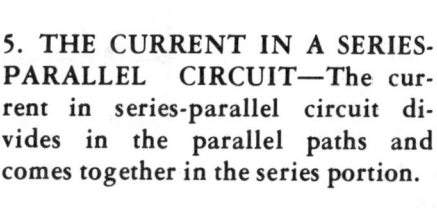

5. THE CURRENT IN A SERIES-PARALLEL CIRCUIT—The current in series-parallel circuit divides in the parallel paths and comes together in the series portion.

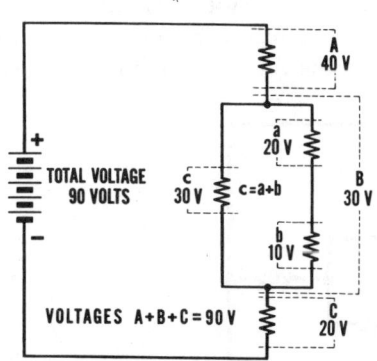

6. THE VOLTAGE IN A SERIES-PARALLEL CIRCUIT—The voltages in series-parallel circuits divide up so that the sum of the voltages in the series portion and the parallel portion is equal to the total voltage.

Self-Test— Review Questions

1. What is the equivalent resistance of the following circuit?

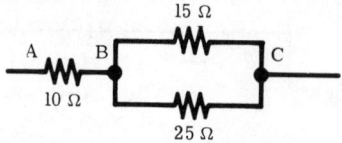

2. In the circuit above, how much current would flow if a voltage of 12 volts were connected across points AC?
3. In the circuit of question 1, with 12 volts applied, what is the voltage between AB? Between BC? Between AC?
4. What is the equivalent resistance of the following circuits?

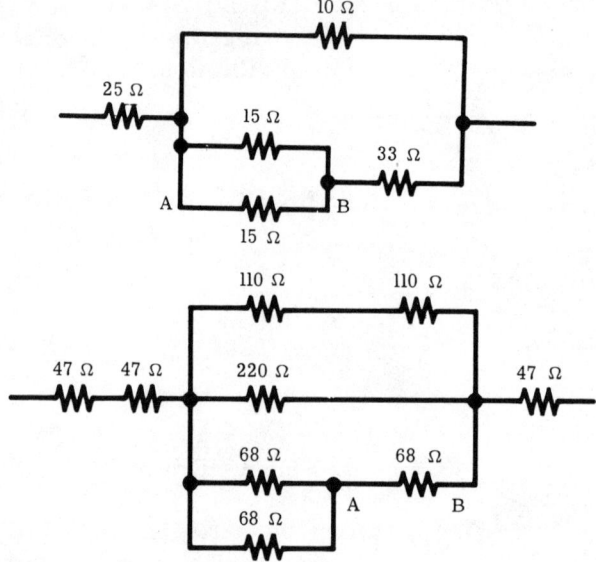

5. In the circuits of question 4, find the voltage across points AB for each circuit. Assume that the total voltage across the circuit is 24 volts.
6. Solve for all the voltages and currents in the following circuit. Do the results conform to Kirchhoff's Laws? Show the direction of current flow.

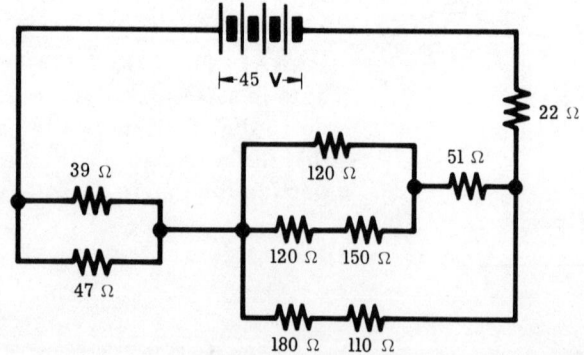

DC SERIES-PARALLEL CIRCUITS

Experiment/Application—Series-Parallel Connections

Suppose that three 30-ohm resistors are connected together, with one resistor in series with a parallel combination of the other two, thus forming a series-parallel circuit. The total resistance is found by combining the parallel 30-ohm resistors to obtain their equivalent resistance, which is 15 ohms, and by adding this value to the 30-ohm resistor in series—making a total resistance of 45 ohms. If the resistors are checked with an ohmmeter, it will read 45 ohms across the entire circuit.

Next, suppose two 30-ohm resistors are connected in series, and a third resistor of the same value is connected in parallel across the series combination. The total resistance can be found by adding the two resistors in the series branch, to obtain an equivalent value of 60 ohms. This value is in parallel with the third 30-ohm resistor, and combining them results in a value of 20 ohms for the total resistance. If you check this value with an ohmmeter, you will see that the meter reading is, indeed, 20 ohms.

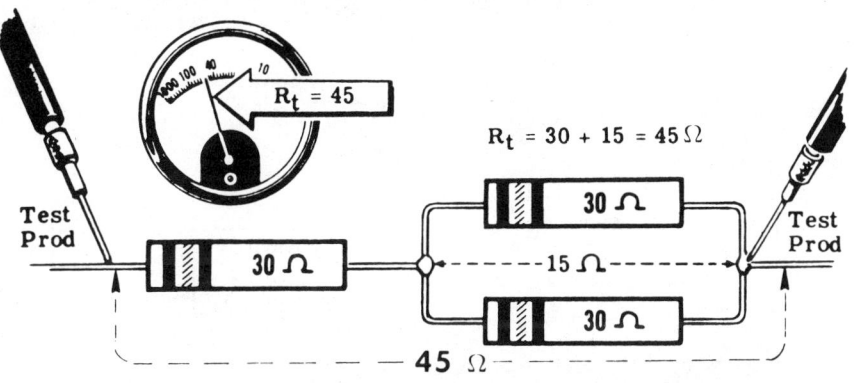

HOW DIFFERENT SERIES-PARALLEL CONNECTIONS .... AFFECT RESISTANCE

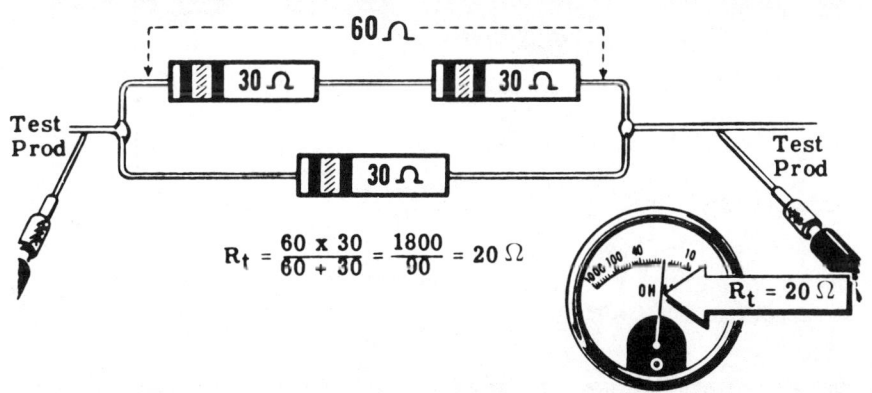

DC SERIES-PARALLEL CIRCUITS

Experiment/Application—Current in Series-Parallel Circuits

Next, consider a series-parallel circuit consisting of two 30-ohm resistors connected in parallel and a 15-ohm resistor in series with one end of the parallel resistors across a 6-volt dry cell battery.

If you connected an ammeter in series with each resistor, in turn, to find the current flow through each, you would see that the current for the 15-ohm series resistor is 0.2 ampere, as is the current at each battery terminal; the current through the 30-ohm resistors, however, is 0.1 ampere each.

Now, suppose the circuit connections are changed so that the 15-ohm and one 30-ohm resistor form a series-connected branch in parallel with the other 30-ohm resistor. An ammeter would now show that the battery current is 0.33 ampere, the 30-ohm resistor current is 0.2 ampere, and the current through the series branch is 0.13 ampere.

SEEING HOW THE CURRENT FLOWS THROUGH SERIES-PARALLEL CIRCUITS

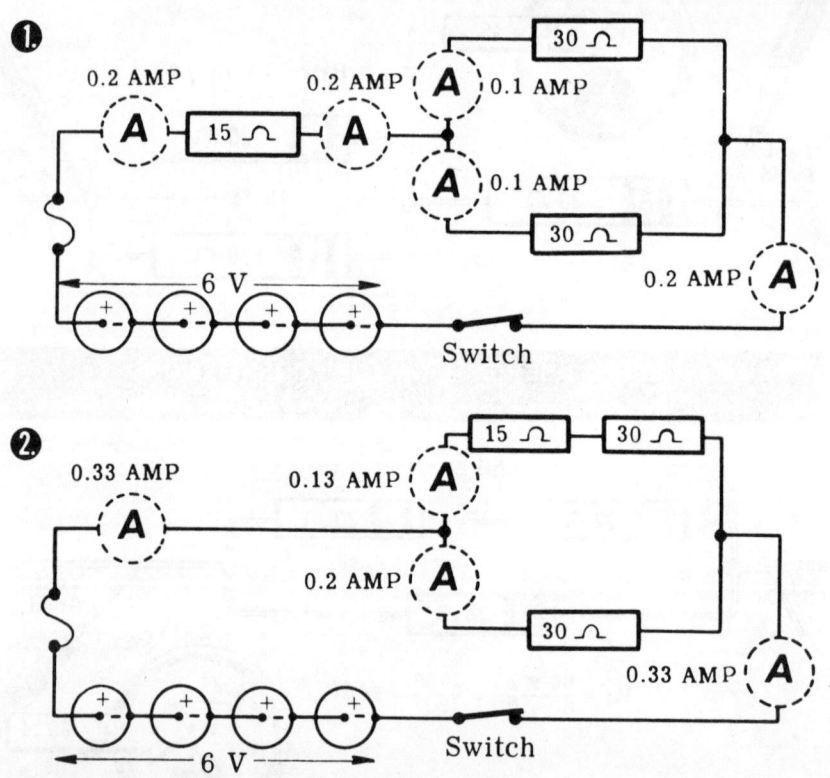

Experiment/Application—Voltage in Series-Parallel Circuits

The division of voltage across series-parallel circuits can be shown by connecting several resistors to form a complex circuit having more than one complete path between the battery terminals, as diagrammed below. As you trace several possible paths across the circuit and measure the voltage across each resistance, you can see that—regardless of the path chosen—the sum of the voltages for any one path always equals the battery voltage. Also, you see that the voltage drop across resistors of equal value differs, depending on whether they are in a series or parallel part of the circuit, and depending also on the total resistance of the path in which they are located.

SEEING HOW VOLTAGE DIVIDES IN A SERIES-PARALLEL CIRCUIT

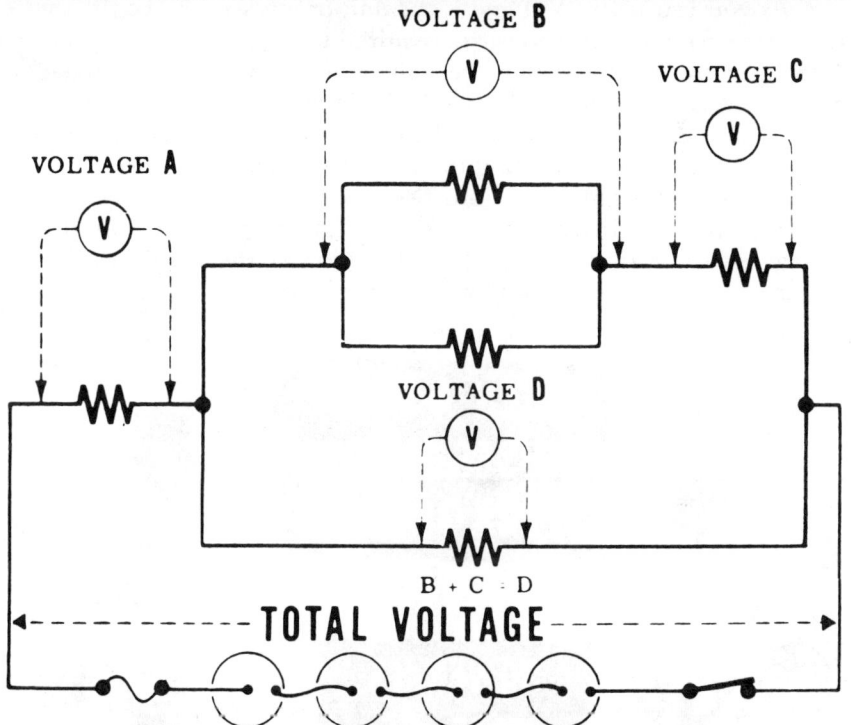

VOLTAGES A + D = TOTAL VOLTAGE
VOLTAGES A + B + C = TOTAL VOLTAGE

Learning Objectives—Next Section

Overview—Now that you know how to solve all kinds of dc circuits, you will learn about electric power. You will learn how it is calculated and how to find the power in dc circuits.

What Electric Power Is

As you know, whenever a force of any kind causes motion, *work* is said to be done. When a mechanical force, for instance, is used to lift a weight, work is done. A force which is exerted *without* causing motion—for example, such as the force of a spring held under tension between two objects which do *not move*—that force does *not* cause work to be done.

You also know that a difference in potential between any two points in an electric circuit gives rise to a voltage which, when the two points are connected, causes electrons to move and current to flow. Here is an obvious case of a force causing motion, and thus of causing work to be done. Whenever a voltage causes electrons to move, therefore, work is done in moving them.

As you learned in Volume 1, the unit of work in the English system is the foot-pound, i.e., the energy required to *raise* 1 pound a distance of 1 foot. In the metric system the unit of work is based on meters and grams and is called the *joule*, with 1 joule equal to about 3/4 of a foot-pound. We can get work back by letting the weight of 1 pound *fall* a distance of 1 foot after connecting the weight to something to get an output of work as the weight falls. Thus, we can draw an analogy between the lifting of the weight and the generation of the potential difference or voltage at a power station, and the lowering of the weight to do work and the flow of electrons to do work at the load end.

The *rate* at which the work of moving electrons from point to point is done is called *electric power*. It is represented by the symbol P, and the unit of power is the *watt*, usually represented by the symbol W. *The watt can be practically defined as the rate at which work is being done in a circuit in which a current of 1 ampere is flowing when the voltage applied is 1 volt.*

The Power Formula

As you learned in Volume 1, it is the ease with which electric power can be transmitted from place to place, and converted to other forms of energy, that makes it so valuable. For example, electrical energy can be converted to heat, light, or accoustical and mechanical energy. The rate of energy conversion is what the engineer really means by the word *power*.

The rate at which work is done in moving electrons through a resistor obviously depends on how many electrons there are to be moved. In other words, the *power consumed in a resistor is determined by the voltage measured across it, multiplied by the current flowing through it.* Expressed in units of measure, this becomes

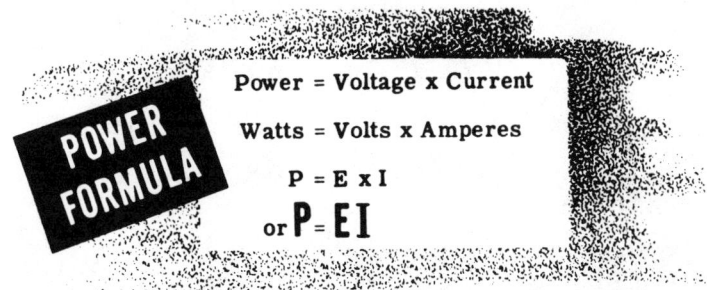

POWER FORMULA

Power = Voltage x Current

Watts = Volts x Amperes

$P = E \times I$

or $P = EI$

In the circuit below, a 15-ohm resistor is connected across a supply of 45 volts. How much power is used up when a current of 3 amperes flows through the resistor?

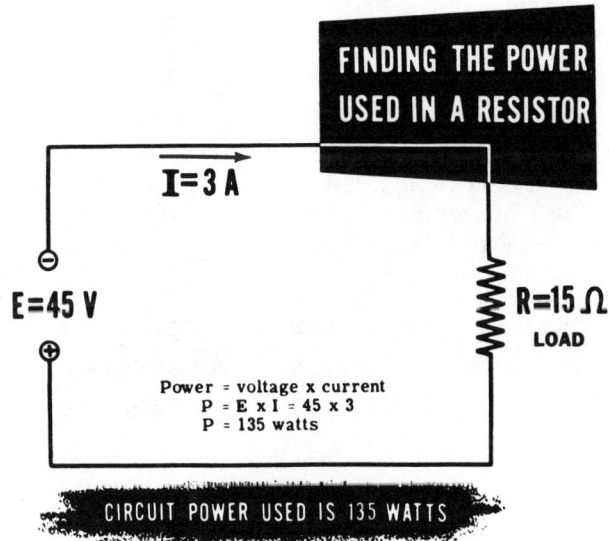

FINDING THE POWER USED IN A RESISTOR

$I = 3 A$

$E = 45 V$

$R = 15 \Omega$

LOAD

Power = voltage x current
$P = E \times I = 45 \times 3$
$P = 135$ watts

CIRCUIT POWER USED IS 135 WATTS

For direct-current (dc) circuits, you can always find the power in a circuit by using the *power formula*.

The Power Formula (continued)

The power formula you learned on the last page, P = EI, can obviously be expressed alternatively in terms of current and resistance, or of voltage and resistance, by the use of our old friend, Ohm's Law. (As E or V can be used interchangeably, this formula can also be stated as P = VI. We will use E.) Since E = IR, the E in the power formula can be replaced by its equal value of IR, and the power used can be calculated *without* the voltage being known.

VARIATION OF THE POWER FORMULA

$$P = EI$$

SUBSTITUTING (IR) FOR E: $P = (IR)I$ OR $I \times R \times I$

SINCE I x I IS I²: $P = I^2 R$

Equally, of course, $I = \frac{E}{R}$. So if E is substituted in the power formula for I, the power used can be found with only the voltage and the resistance being known.

ANOTHER VARIATION

$$P = EI$$

SUBSTITUTING $\frac{E}{R}$ FOR I: $P = E\left(\frac{E}{R}\right)$ OR $\frac{E \times E}{R}$

SINCE E x E IS E²: $P = \frac{E^2}{R}$

The Conversion Tables you learned in Volume 1 apply to the watt just as they do to the volt, the ampere, the ohm, and all the others. Quantities of power greater than 1,000 watts are generally expressed in kilowatts (kW), and quantities greater than 1,000,000 watts are generally expressed as megawatts (MW). Quantities less than 1 watt are generally expressed in milliwatts (mW).

LARGE AND SMALL UNITS OF POWER

1 megawatt = 1,000,000 watt = 1,000,000 W = 1 MW
1 kilowatt = 1000 watts
1 kw = 1000 W
1 milliwatt = $\frac{1}{1000}$ watt
1 mw = $\frac{1}{1000}$ w
1 microwatt = $\frac{1}{1,000,000}$ watt = $\frac{1}{1,000,000}$ W = 1 μW

ELECTRIC POWER

Power Rating of Equipment

You have probably found from your own experience that most electrical equipment is rated for both voltage and power—volts and watts. Electric lamps rated at 120 volts for use on 120-volt lines are also rated in watts, and are usually identified by wattage rather than by voltage.

Perhaps you have wondered what this rating in watts means and indicates. *The wattage rating of an electric lamp or other electrical equipment indicates the rate at which electrical energy is changed into another form of energy, such as heat or light.* The faster a lamp changes electrical energy to light, the brighter the lamp will be; thus, a 100-watt lamp furnishes more light than a 75-watt lamp.

Electric soldering irons are made in various wattage ratings, with the higher wattage irons changing electrical energy to heat faster than those of a lower wattage rating. Similarly, the wattage rating of motors, resistors, and other electrical devices indicates the rate at which they are designed to change electrical energy into some other form of energy. Motors are often classified in terms of horsepower—which you will learn more about later. Horsepower is another unit for measuring the rate at which work is done, and 1 horsepower is equal to 746 watts. You will learn more about horsepower when we study motors.

Power rating of equipment is the rate at which it changes electrical energy into . . . .

HEAT

75 WATTS

150 WATTS

or

LIGHT

Greater wattage furnishes more heat and light.

Power Rating of Equipment (continued)

When power is used in a material having resistance, electrical energy is changed into heat. When more power is used in the material, the rate at which electrical energy is changed to heat increases, and the temperature of the material rises. If the temperature rises too high, the material may change its composition, expand, contract, or burn. For that reason, all types of electrical equipment are rated for a maximum wattage. This rating may be in terms of watts, or in terms of maximum voltage and current—which effectively gives the rating in watts.

Resistors are rated in watts as well as in ohms of resistance. Resistors of the same resistance value are available in different wattage values. Carbon resistors, for example, are commonly made in wattage ratings of 1/4, 1/2, 1, and 2 watts. The larger the size of carbon resistor, the higher its wattage rating, since a larger amount of material will absorb and give up heat more easily.

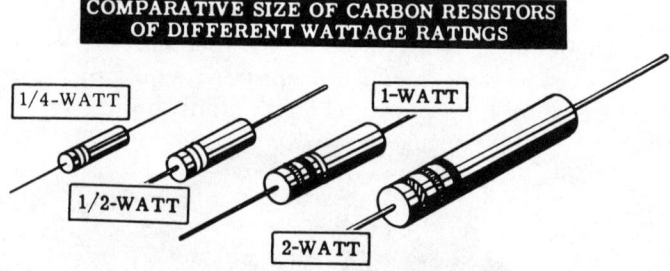
COMPARATIVE SIZE OF CARBON RESISTORS OF DIFFERENT WATTAGE RATINGS

When resistors of wattage ratings greater than 2 watts are needed, wire-wound resistors are used. Such resistors are made in ranges between 5 and 200 watts, with special types being used for power in excess of 200 watts.

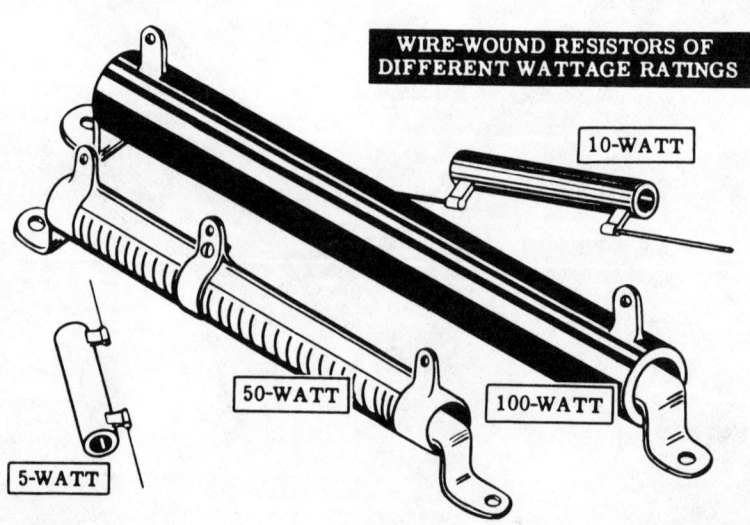
WIRE-WOUND RESISTORS OF DIFFERENT WATTAGE RATINGS

Fuses

You know that when current passes through a resistor, electrical energy is transformed into heat, which raises the temperature of the resistor. If the temperature rises too high, the resistor may be damaged. The metal wire in a wound resistor may melt, thereby opening the circuit and interrupting current flow. This effect is used to advantage in fuses.

Fuses are resistors using special metals with very low resistance values and a low melting point, which are designed to *blow out* and thus open the circuit when the current exceeds the fuse's rated value. When the power consumed by the fuse raises the temperature of the metal too high, the metal melts and the fuse *blows*. Blown fuses can usually be identified by a broken filament and darkened glass. If you are uncertain, you can remove the fuse and check it with an ohmmeter.

You have already learned that excessive current may seriously damage electrical equipment—motors, test instruments, radio receivers, etc. Fuses are cheap, yet the other equipment is much more expensive.

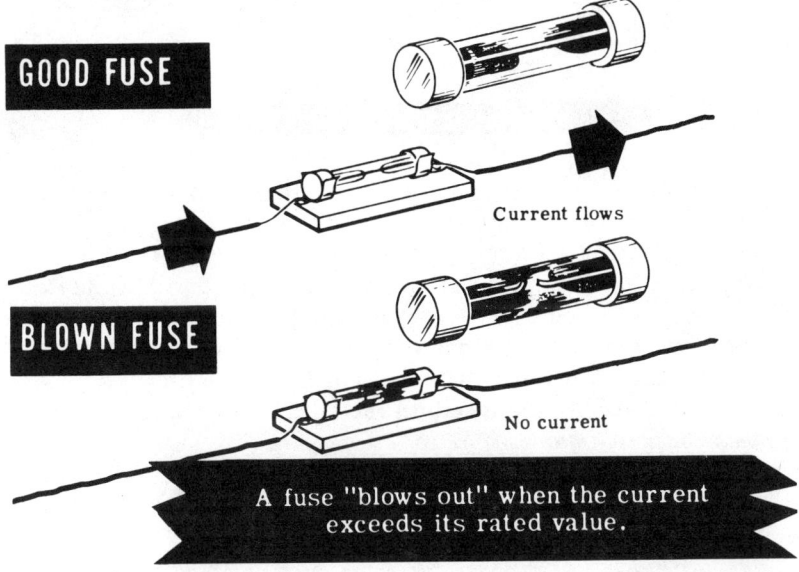

A fuse "blows out" when the current exceeds its rated value.

There are two types of fuses in use today—*conventional* fuses, which blow immediately when the circuit is overloaded, and *slow-blowing* fuses. Slow-blowing (slo-blo) fuses can accept momentary overloads without blowing, but if the overload continues, they will open the circuit. These slo-blo fuses are used in circuits that have a sudden rush of high current when turned on, such as motors and some appliances. If such circuits used a conventional fuse with a high enough value to handle the high starting currents, there would be little protection under normal running conditions. It is important that you replace fuses with the proper type, whether conventional or slow blowing.

Fuses (continued)

Although it is the power used by a fuse which causes it to blow, fuses are rated by the *current* which they will conduct without burning out, since it is high current which damages equipment. Since various types of equipment use different currents, fuses are made in many sizes, shapes, and current ratings.

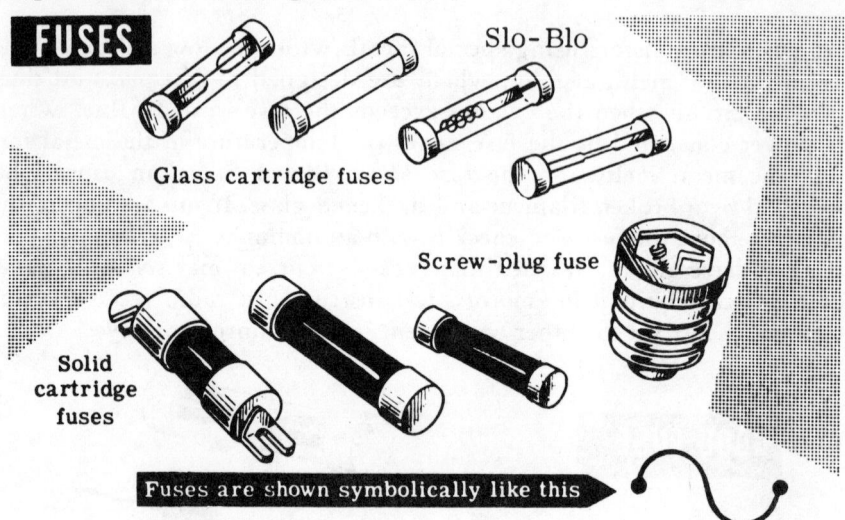

It is important that you always use fuses with the proper current rating—slightly higher than the greatest current you expect in the circuit. Too low a rating will result in unnecessary blowouts, while too high a rating may allow dangerously high currents to pass. In the experiments to follow, the circuits will be *fused* to protect the ammeter. Since the range of the ammeter is 0 to 1 ampere, a 1.5-ampere fuse will be used.

The fuse is inserted in the circuit by connecting the fuse holder in series and snapping the fuse into the holder—but <u>*always remember to disconnect the power source before you change a fuse*</u>!

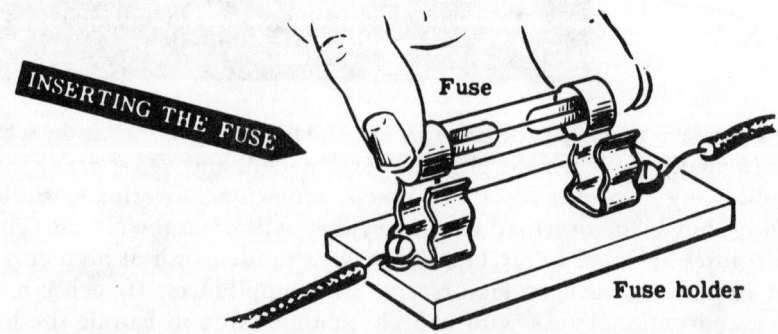

Later, you will learn about another protective device called the *circuit breaker* that provides protection without the inconvenience of changing fuses.

Power in Series Circuits

The total power consumed in a series circuit is the sum of the power used in all the individual circuit elements and is easily found.

Consider a circuit in which three resistors—R1 = 20 ohms, R2 = 16 ohms, and R3 = 12 ohms—are connected in series across a power supply of 72 volts. You now need to know how much power will be consumed by the circuit.

First, sketch the circuit diagram, and fill in the known values.

Then calculate circuit current, which you can do as soon as you have the circuit resistance, R_t. Here you see that R_t = 20 + 16 + 12 = 48 ohms.

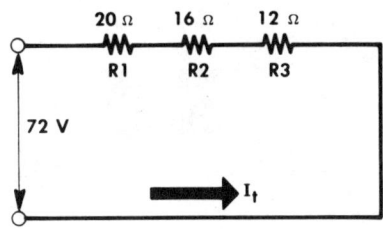

With the voltage and circuit resistance both known, Ohm's Law tells you that the circuit current is:

$$I_t = \frac{E}{R} = \frac{72}{48} = 1.5 \text{ amperes}$$

You can now use the variant of the power formula which gives you P when you know only I and R. It is, you will remember, $P = I^2R$.

$$P1 = I^2R1 = 1.5 \times 1.5 \times 20 = 45 \text{ watts}$$
$$P2 = I^2R2 = 1.5 \times 1.5 \times 16 = 36 \text{ watts}$$
$$P3 = I^2R3 = 1.5 \times 1.5 \times 12 = 27 \text{ watts}$$

Since the power taken by a series circuit is the sum of the power taken by the individual resistors in the circuit, you find that

$$P_t = 45 + 36 + 27 = 108 \text{ watts}$$

And, since Ohm's Law tells us that P = EI—and P = 72 volts and I = 1.5 amperes, then

$$P = 72 \times 1.5 = 108 \text{ watts}$$

Another way of attacking this problem would be to simplify the circuit before you start to calculate I, and to draw the equivalent circuit like this.

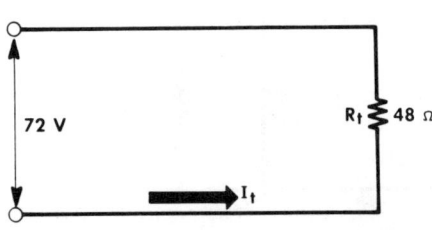

Now calculate I_t by using Ohm's Law exactly as you did before:

$$I_t = \frac{E}{R_t} = \frac{72}{48} = 1.5 \text{ amperes}$$

Then the same variant of the power formula gives you the power consumed by the circuit:

$$P_t = I^2R_t = 1.5 \times 1.5 \times 48 = 108 \text{ watts}$$

which is, as you would expect, the same answer as you found before.

Power in Parallel Circuits

You have seen that the total power taken by a series circuit is equal to the sum of the power taken by all the individual resistors in the circuit.

The same thing is true of all parallel circuits. *The total power taken by a parallel circuit is equal to the sum of the power taken by all the individual resistors in the circuit.* It can be found by multiplying the total circuit current by the voltage across the circuit.

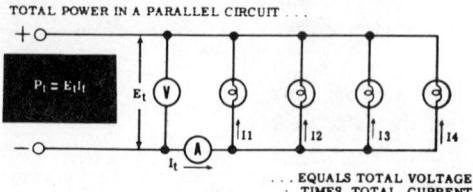

TOTAL POWER IN A PARALLEL CIRCUIT ...
... EQUALS TOTAL VOLTAGE TIMES TOTAL CURRENT

Total power = 100 + 100 + 100 + 100 = 400 watts

By the power formula

$$P_t = E_t \times I_t = 125 \times 3.2 = 400 \text{ watts}$$

If either circuit current or circuit voltage is unknown, circuit power can still be found by applying the rules for parallel circuits to calculate total circuit resistance—but only, of course, if the value of every individual resistor in the circuit is known.

It is then only a matter of selecting the correct variation of the power formula to take advantage of the particular set of facts you already know. The two possibilities are shown in the double diagram below.

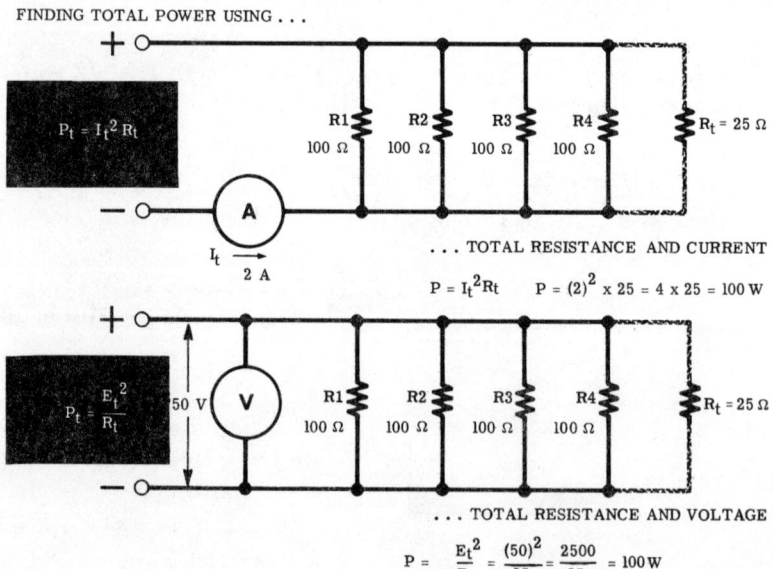

FINDING TOTAL POWER USING ...

... TOTAL RESISTANCE AND CURRENT

$P = I_t^2 R_t \qquad P = (2)^2 \times 25 = 4 \times 25 = 100 \text{ W}$

... TOTAL RESISTANCE AND VOLTAGE

$$P = \frac{E_t^2}{R_t} = \frac{(50)^2}{25} = \frac{2500}{25} = 100 \text{ W}$$

ELECTRIC POWER

Power in Complex Circuits

You have seen that the power taken in both series and parallel circuits is equal to the sum of the power taken by all the individual resistors or loads.

The same thing is true of complex circuits involving series and parallel parts. The *total power taken by a series-parallel circuit is equal to the sum of the power taken by all the individual loads in the circuit.* It can be found by multiplying the total current by the voltage across the circuit.

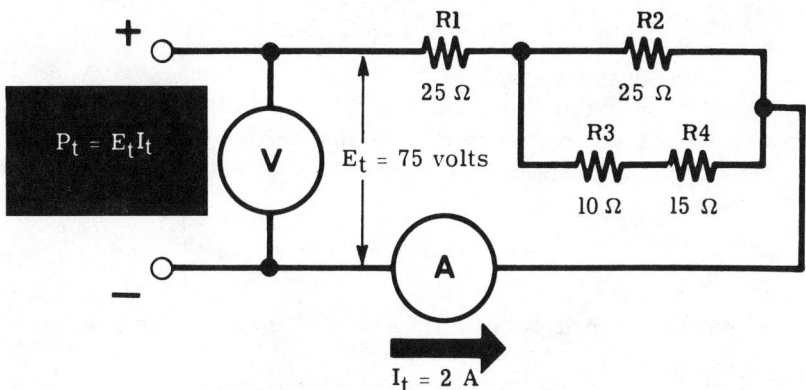

TOTAL POWER = $P_t = E_t I_t$ = 75 x 2 = 150 watts

Since we know that the total circuit current is 2 amperes, and this current must flow through R1, then using the right version of the power formula for current and resistance will give us

$$P_{R1} = I_t^2 \times R1 = (2)^2 \times 25 = 100 \text{ watts}$$

Since the total wattage is 150 watts, then the total wattage in R2, R3, and R4 must be 50 watts (150 − 100 = 50). Let us check this by combining R2, R3, and R4 into their equivalent resistance.

$$R_a = R3 + R4 = 10 + 15 = 25 \text{ ohms}$$

The circuit is now

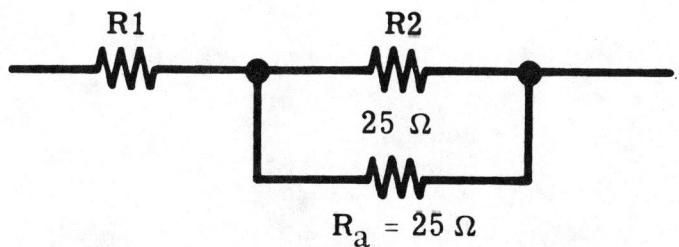

The parallel combination of R_a and R2 is equal to

$$R_b = \frac{R2 \times R_a}{R2 + R_a} = \frac{25 \times 25}{25 + 25} = 12.5 \text{ ohms}$$

Power in Complex Circuits (continued)

The total current is now

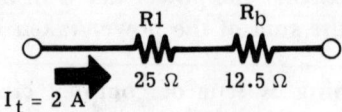

The power in R_b (parallel combination) can be calculated in a number of ways. The simplest way is to use the power formula for current and resistance,

$$P_{Rb} = I_t^2 \times R_b = (2)^2 \times 12.5 = 50 \text{ watts}$$

which is what we expected.

To find the wattage dissipated in R2, R3, and R4 we can use some facts that we learned earlier. We know that the equivalent resistance R_b is 12.5 ohms and that the total current is 2 amperes. Therefore, using Ohm's Law,

$$E_b = I_t \times R_b = 2 \times 12.5 = 25 \text{ volts}$$

Therefore, the power in R2 can be calculated from the power formula.

$$P_{R2} = \frac{E^2}{R} = \frac{(25)^2}{25} = 25 \text{ watts}$$

Since R_a is 25 watts, then R3 and R4 together dissipate 25 watts also. The current in the branch of the parallel circuit containing R3 and R4 can be calculated by Ohm's Law as

$$I_{R3+R4} = \frac{E_{R3+R4}}{R_{R3 \times R4}} = \frac{25}{25} = 1 \text{ ampere}$$

The power in R3 is

$$P_{R3} = (I_{R3})^2 \times R3 = (1)^2 \times 10 = 10 \text{ watts}$$

The power in R4 is

$$P_{R4} = (I_{R4})^2 \times R4 = (1)^2 \times 15 = 15 \text{ watts}$$

The total power is the sum of the power in each resistor.

Power of R1 = 100 watts
Power of R2 = 25 watts
Power of R3 = 10 watts
Power of R4 = 15 watts
Total Power = 150 watts

Thus, again, it is clear that no matter how complex the circuit, the total power is the sum of the individual wattage and is equal to the input voltage times the input current.

ELECTRIC POWER

Review of Electric Power

Whenever an electric current flows, work is done in moving electrons through the conductor. The electrons to be moved may be moved in either a short or a long period of time, and the rate at which the work is done is called electric power.

1. ELECTRIC POWER—The rate of doing work in moving electrons through a material is electric power, "P." The basic unit of power is the watt, "W." 1 watt is used to make a current of 1 ampere flow through a resistance of 1 ohm.

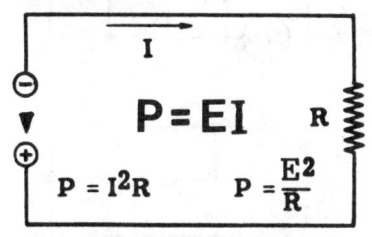

$$P = EI$$
$$P = I^2 R \qquad P = \frac{E^2}{R}$$

2. POWER FORMULA—Electric power used in a resistance equals the voltage across the resistance terminals times the current flow through the resistance. It is also equal to the current squared, times the resistance, or to the voltage squared, divided by the resistance.

75 Watts

150 Watts

3. POWER RATING—Electrical equipment is rated according to the rate at which it uses electric power. The power used is converted from electrical energy into some other form of energy, such as heat or light.

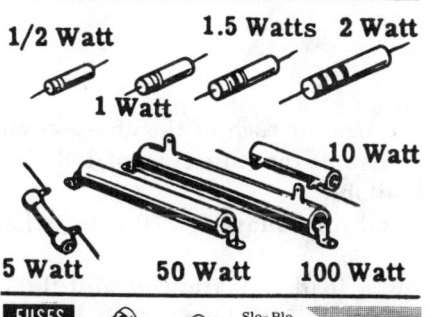

1/2 Watt   1.5 Watts   2 Watt
1 Watt
10 Watt
5 Watt   50 Watt   100 Watt

4. RESISTOR POWER RATINGS—Resistors are rated both in ohms of resistance and by reference to the maximum power which can safely be used in the resistor. High-wattage resistors are constructed larger than low-wattage resistors, so as to provide a greater surface for dissipating heat. They are also made of materials that can withstand greater quantities of heat.

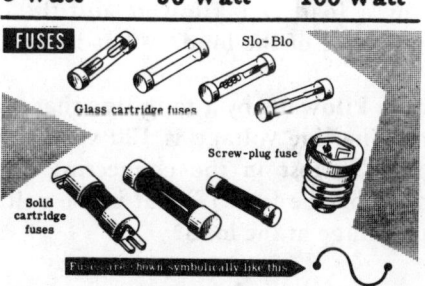

FUSES
Glass cartridge fuses
Slo-Blo
Screw-plug fuse
Solid cartridge fuses
Fuses are shown symbolically like this

5. FUSES—Fuses are metal resistors designed to open an electric circuit if the current through them exceeds their rated value.

Self-Test— Review Questions

1. Calculate the power in the circuits shown below:

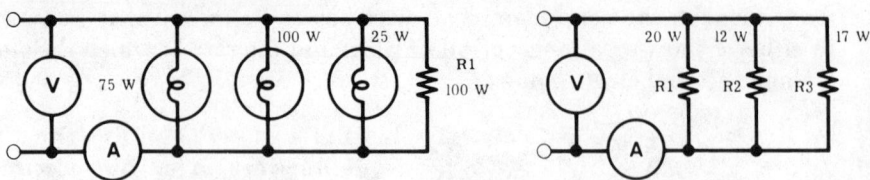

2. Calculate the power—total and in each circuit element—for the circuits shown below:

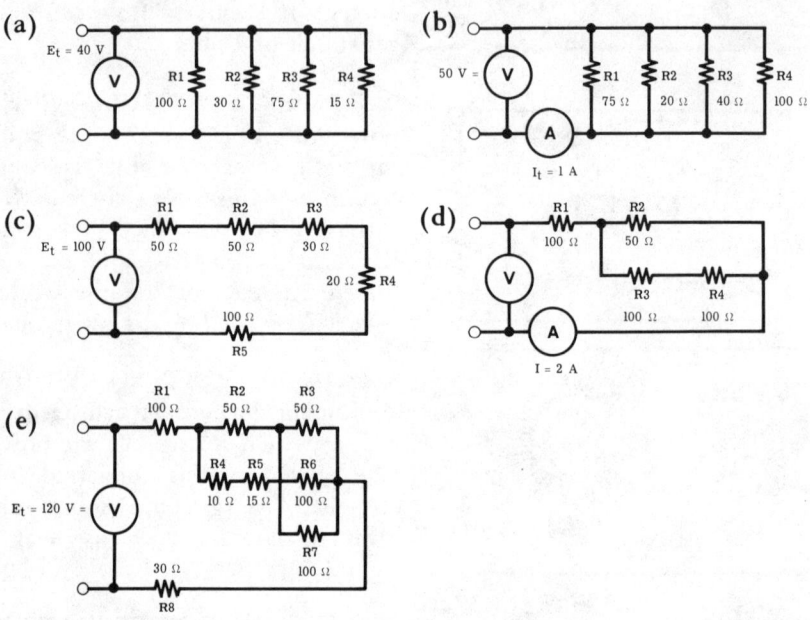

3. Calculate and specify the proper fuse size for each of the above circuits, assuring about a 50% safety factor. Assume that fuse sizes of 1/8, ¼, ½, 1, 2, 3, 5, 7.5, and 10 amperes are available.
4. If the power in a line is 200 watts and the voltage is 120 volts, what is the line current?
5. You know that the power drawn by a load is 1 kilowatt and the line voltage is 120 volts. What is the resistance of the load? What is the line current?
6. You must connect a load that draws 5 kilowatts by a long line that has a resistance (in each leg) of 0.5 ohm. The line voltage is 120 volts at the input to the line. How much power is lost in the connecting lines? (Neglect the effect of the voltage drop on the load.) What is the voltage drop in the line? What is the actual voltage at the load?

ELECTRIC POWER

Experiment/Application—The Use of Fuses

You have seen how a resistor overheats when it uses more power than its power rating. Now you will see how this effect is put to use to protect electrical equipment from damage due to excessive currents.

Suppose you connect four dry cells in series to form a 6-volt battery and then connect a 15-ohm, 10-watt resistor, a knife switch, a fuse holder, and an ammeter in series across the battery, and insert a 1/8-ampere fuse in the fuse holder. When you close the switch, the fuse will *blow*, thus opening the circuit so that no current can flow, as shown by a zero reading of the ammeter.

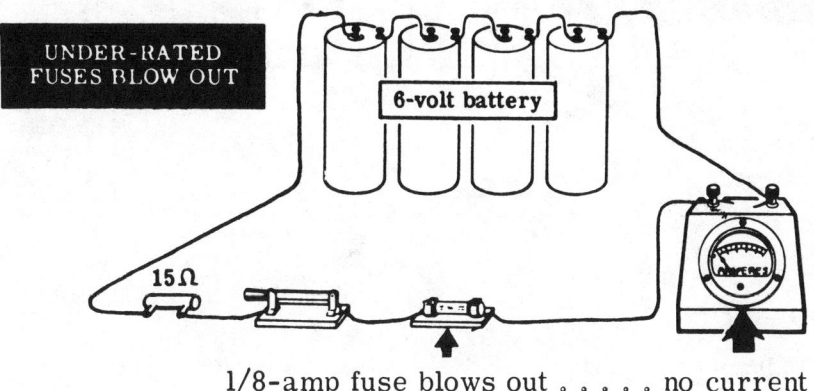

1/8-amp fuse blows out . . . . . <u>no current</u>

However, if you insert a 1/2-ampere fuse in the fuse holder, it will not blow, and the ammeter will show a current flow.

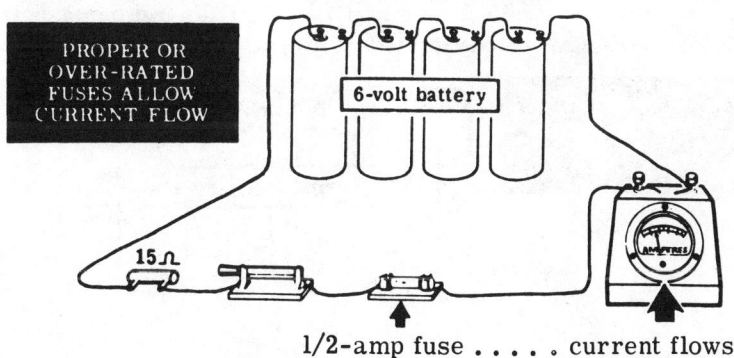

1/2-amp fuse . . . . . <u>current flows</u>

Since the resistance of the circuit is 15 ohms and the voltage is 6 volts, the current flow by Ohm's Law is about 0.4 ampere ($\frac{6 \text{ volts}}{15 \text{ ohms}}$). The 1/8-ampere (0.125-ampere) fuse *blows out* because the current exceeds its rating, and it will not carry 0.4 ampere. However, the 1/2 ampere (0.5 ampere) fuse carries the current without blowing, since its rating exceeds the actual current flow.

Experiment/Application—How Fuses Protect Equipment

Using the circuit shown previously, note that the 15-ohm resistor limits the current through the circuit sufficiently to keep a 1/2-ampere fuse from burning out. The circuit operates without damage to the ammeter.

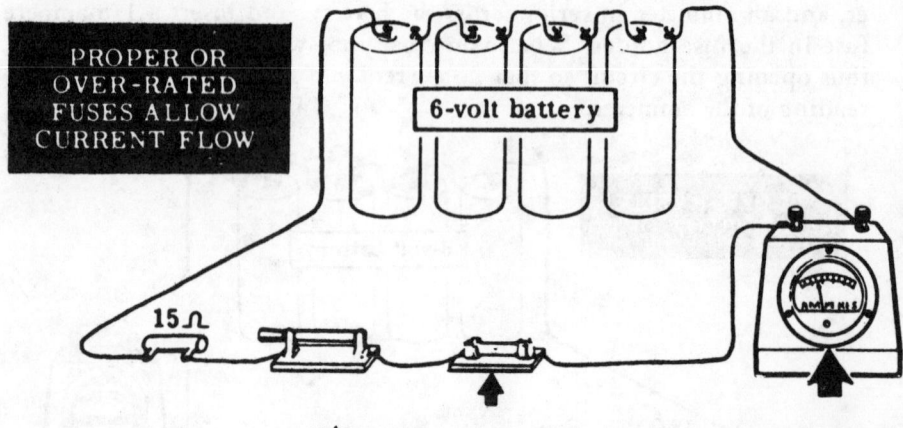

PROPER OR OVER-RATED FUSES ALLOW CURRENT FLOW

1/2-amp fuse . . . . current flows

If you were to short-circuit the resistor as in the diagram below, the fuse would burn out and open the circuit without damage to the ammeter. Because the fuse serves as the *predetermined* weakest link in this circuit, it is the electrical safety device. In choosing fuses, be sure not to choose one whose rating is too high for the expected current flow. If trouble occurs, the highly over-rated fuse may not burn out before the meter does so that all protection is lost for the meter.

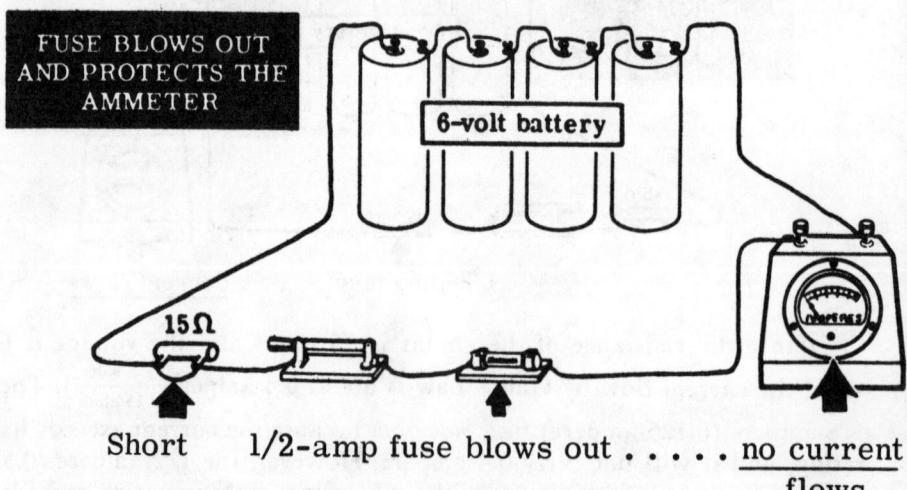

FUSE BLOWS OUT AND PROTECTS THE AMMETER

Short   1/2-amp fuse blows out . . . . . no current flows

ELECTRIC POWER

Experiment/Application—Power in Series Circuits

To show that power can be determined when any two of the circuit variables—current, voltage, and resistance—are known, connect three 15-ohm, 10-watt resistors in series across a 9-volt dry cell battery.

After measuring the voltage across each resistor, you can apply the power formula $P = \dfrac{E^2}{R}$ to find the power for each resistor. You see that the power used by each resistor is about 0.6 watt and that the total power is about 1.8 watts.

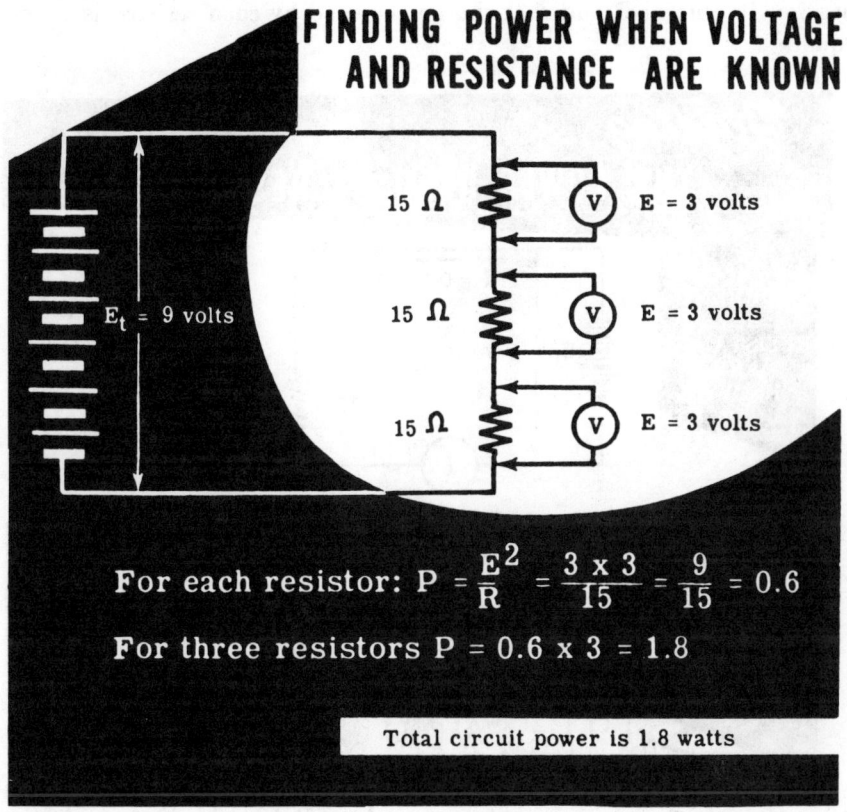

FINDING POWER WHEN VOLTAGE AND RESISTANCE ARE KNOWN

For each resistor: $P = \dfrac{E^2}{R} = \dfrac{3 \times 3}{15} = \dfrac{9}{15} = 0.6$

For three resistors $P = 0.6 \times 3 = 1.8$

Total circuit power is 1.8 watts

To show that the same results are obtained using current and resistance or current and voltage, connect an ammeter in the circuit to measure current. The power used by each resistor is then found by using the power formula in two ways: $P = I^2R$ and $P = EI$. Notice that the power in watts is nearly the same for each variation of the power formula used, with the negligible difference due to meter inaccuracies and slight errors in meter readings.

Experiment/Application—Power in Series Circuits (continued)

To show the effect of the power rating of a resistor on its operation in a circuit, suppose that two 15-ohm resistors—one rated at 10 watts and the other rated at 1 watt—are connected in a series circuit as shown below. The ammeter reads the circuit current and, using the power formula, $P = I^2R$, you find that the power used in each resistor is approximately 1.35 watts. This is slightly more than the power rating of the 1-watt resistor, and you will see that it heats rapidly, while the 10-watt resistor remains relatively cool. To check the power used in each resistor, the voltages across them are measured with a voltmeter and multiplied by the current. Notice that the power is the same as that previously obtained, and that the power used by each resistor is exactly equal.

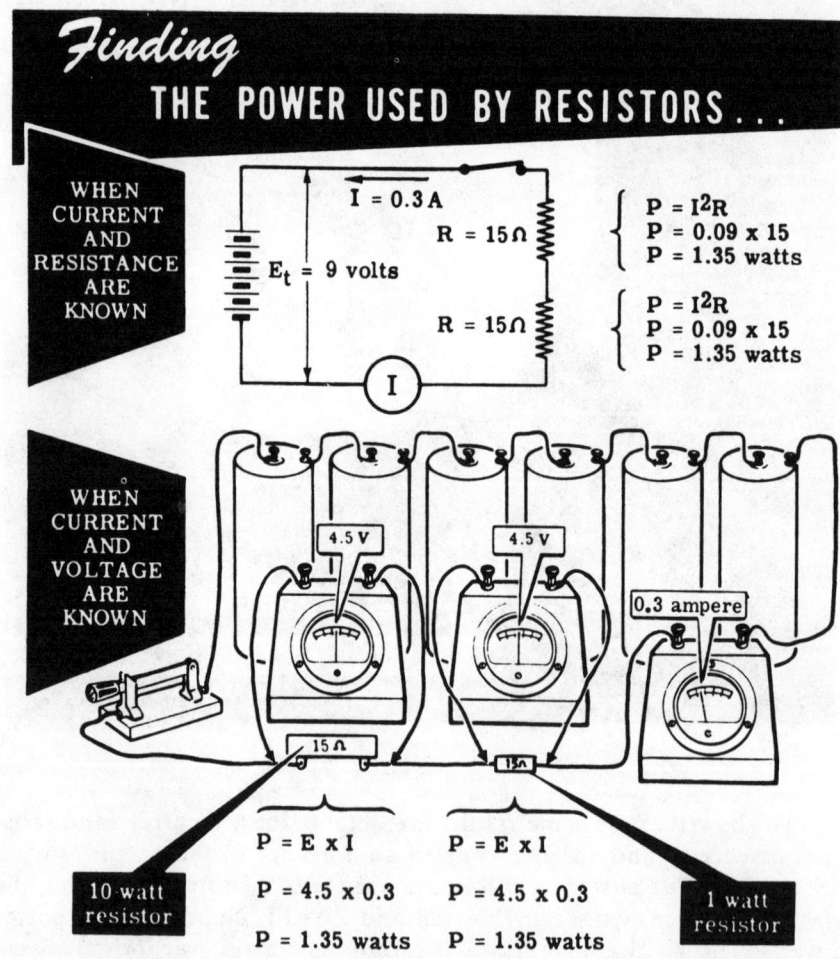

Experiment/Application—Power in Series Circuits (continued)

Next, the 1-watt resistor is replaced by one rated at ½ watt. Observe that it heats more rapidly than the 1-watt resistor and becomes very hot, indicating that the power rating has been greatly exceeded. As the power for each resistor is found (using current and resistance, then voltage and current as a check), you will see that each resistor is using the same amount of power. This shows that the power rating of a resistor does not determine the amount of power used in a resistor. Instead, the power rating only indicates the maximum amount of power that may be used *without damaging* the resistor.

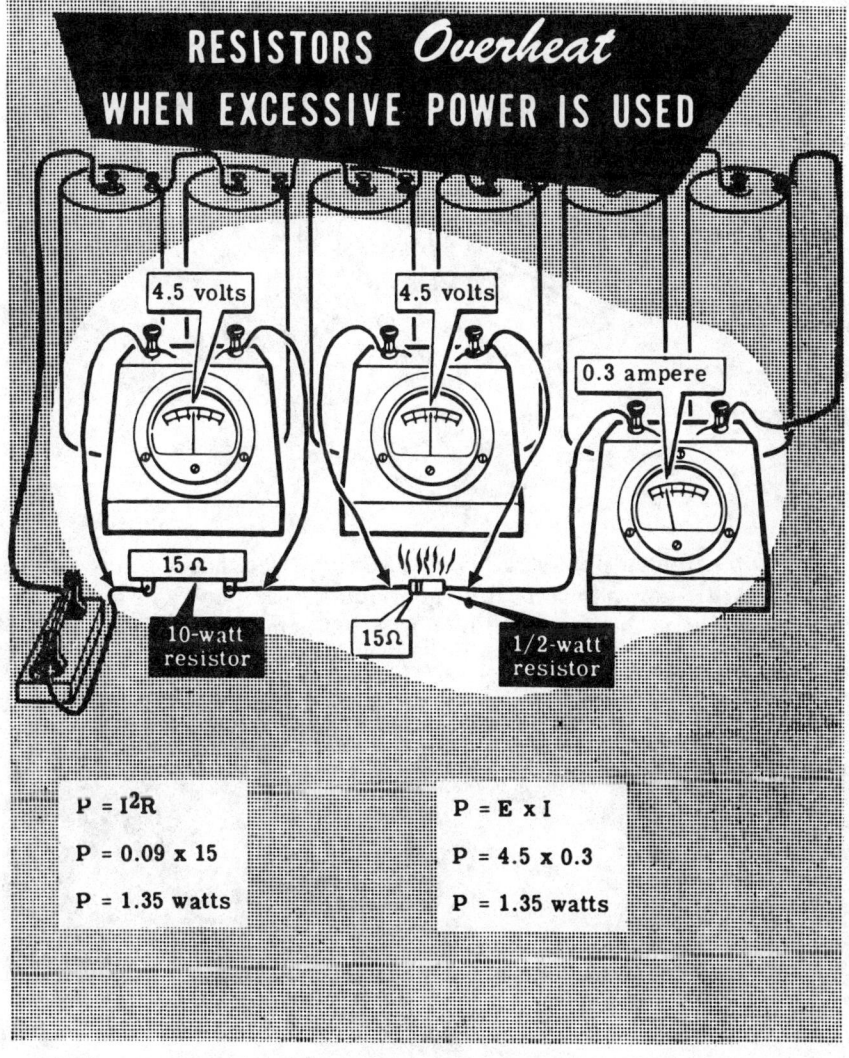

Experiment/Application—Power in Parallel Circuits

To show that the power used by a parallel circuit is equal to the power used by all of the parts of the circuit, connect three lamp sockets in parallel across a 6-volt battery, with a 0-1 ampere range ammeter in series with the battery lead, and a 0-10-volt range voltmeter across the battery terminals. Next, insert 6-volt, 250-milliampere lamps in the lamp sockets, but do not tighten them. When you close the switch, you will see that the voltmeter indicates battery voltage, but the ammeter shows no current flow, since no power is being used by the circuit.

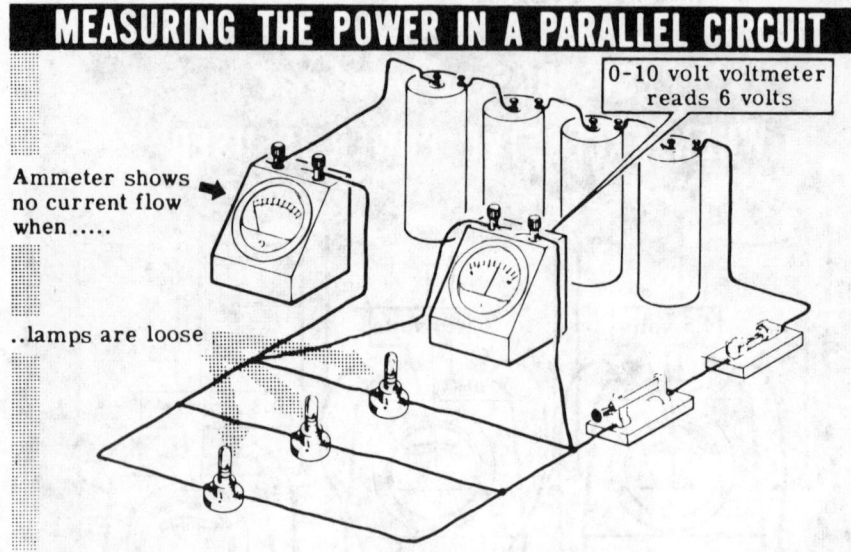

As you tighten one of the lamps, you will see that it lights and the ammeter will show a current flow of about 0.25 ampere. The power used by this one lamp, then, is about 6 volts × 0.25 ampere, or 1.5 watts.

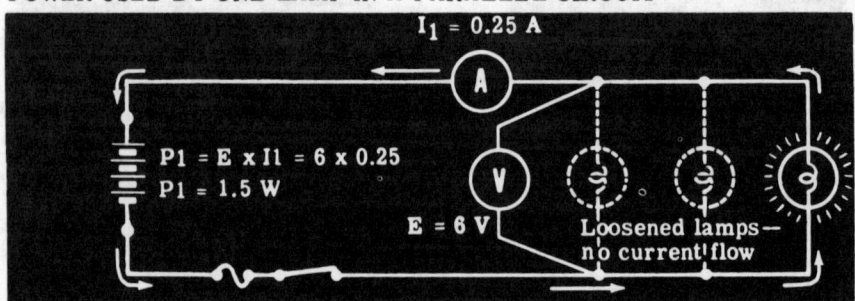

You already know that the voltage across any part of a parallel circuit is equal to the source voltage, so that the voltage across the lamp is equal to the battery voltage.

Experiment/Application—Power in Parallel Circuits (continued)

As you loosen the first lamp and tighten each of the other lamps in turn, you will see that the current—and hence the power used by each lamp—is about the same. The current measured each time is the current through only the one tightened lamp.

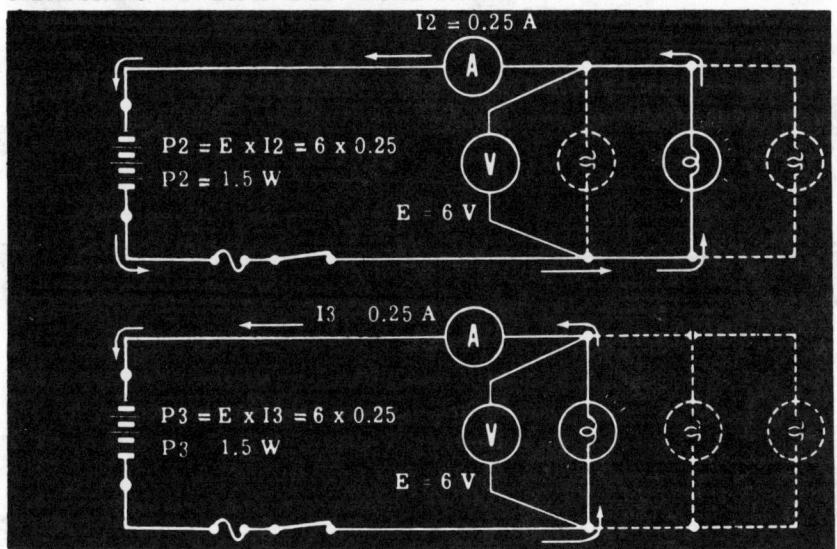

Next, suppose you tighten all three lamps in their sockets. You will see that they all light and that the ammeter shows a circuit current of about 0.75 ampere. The voltage is still about 6 volts, so that the circuit power (P_t) equals 0.75×6, or about 4.5 watts.

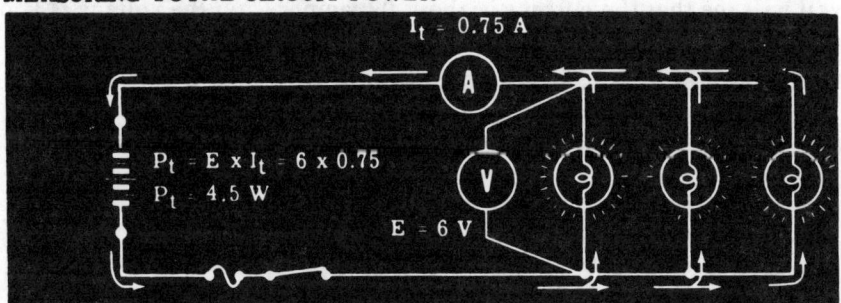

The total circuit power is found to be about 4.5 watts. If you add the power used individually by each of the lamps, the sum is equal to 4.5 watts ($1.5 + 1.5 + 1.5 = 4.5$). Therefore, you can see that the total power used by a parallel circuit is equal to the sum of the power used by each part of the circuit.

Experiment/Application—Power in Parallel Circuits (continued)

Now replace the three lamp sockets with 30-ohm resistors. If you then remove the voltmeter leads from the battery and close the switch, you will notice that the ammeter shows a current flow of about 0.6 ampere.

The total resistance found by applying the rules for parallel circuits is 10 ohms, so that the circuit power is equal to $P = I^2R$ or $(0.6)^2 \times 10 = 3.6$ watts.

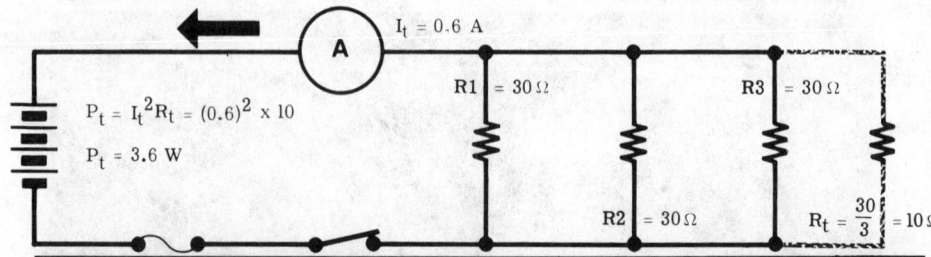

USING TOTAL CURRENT AND RESISTANCE TO MEASURE CIRCUIT POWER

Now remove the ammeter and connect the voltmeter to the battery leads. When the switch is closed, the voltage registers about 6 volts, so that the circuit power is equal to $P = \dfrac{E^2}{R}$ or $\dfrac{(6)^2}{10} = 3.6$ watts.

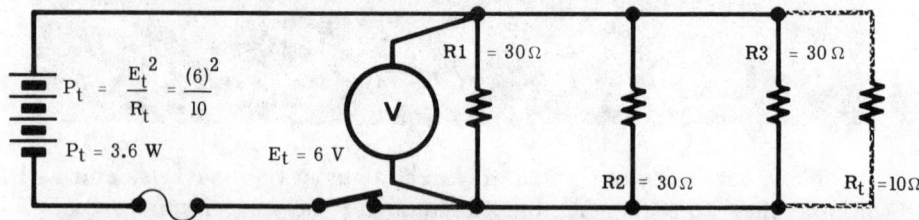

USING TOTAL VOLTAGE AND RESISTANCE TO MEASURE CIRCUIT POWER

Finally, replace the ammeter in the circuit; when power is applied, you will see that the current is about 0.6 ampere and the voltage is about 6 volts, so that the total circuit power is equal to $P = EI$ or $6 \times 0.6 = 3.6$ watts.

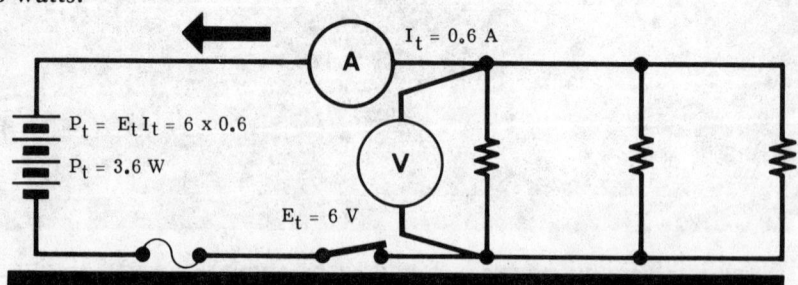

USING TOTAL VOLTAGE AND CURRENT TO MEASURE CIRCUIT POWER

Thus, you see that the total power in a parallel circuit may be determined, as in a series circuit, whenever any *two* of the factors—current, voltage, or resistance—are known.

Thevenin's Theorem—Voltage Division in the Series Circuit (continued from page 2-42)

If you load a voltage divider with an external resistance, the voltage will decrease. It is necessary to take this into consideration sometimes. You could do this by using Ohm's Law and calculating the parallel resistance of R2 and R_{load} and then the voltage divider itself. A simpler method is to use Thevenin's theorem, which enables you to calculate quickly the effect of any load. With Thevenin's theorem you can replace the circuit shown in 1 below (as shown on page 2-41) with that shown in 2A. That is, the voltage source is the old E_{out} without load and the series resistance is the parallel combination of R1 and R2. Illustration 2B shows the Thevenin equivalent circuit.

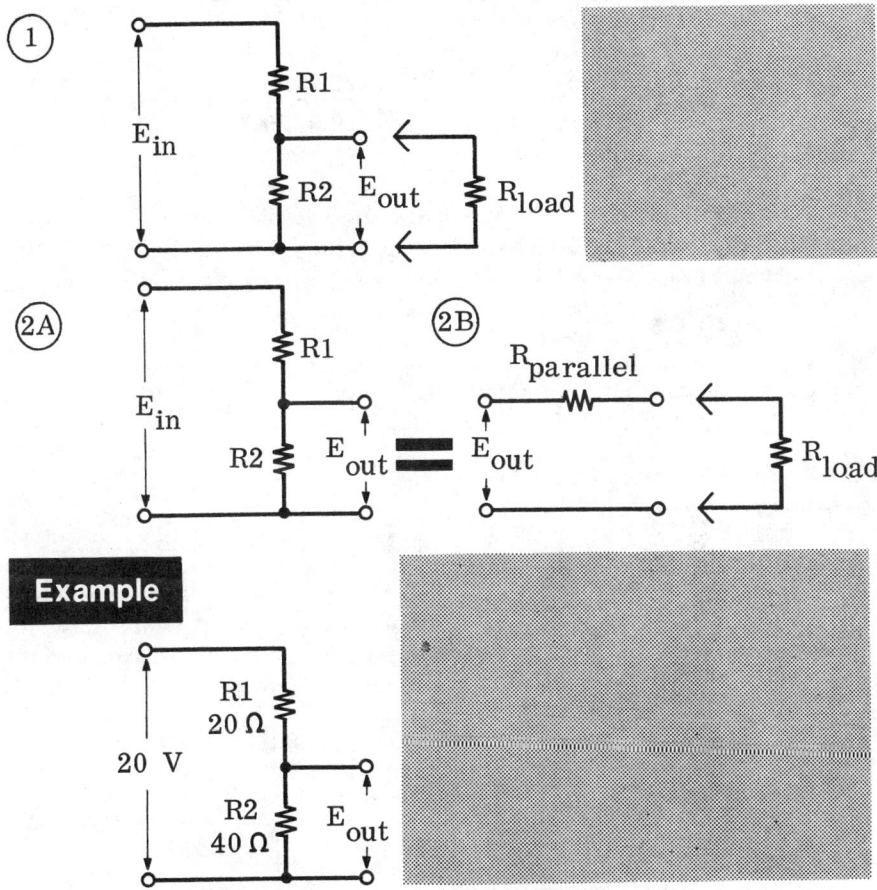

Example

As you know from the voltage divider calculations on pages 2-41 and 2-42, the voltage across R2 with no load is

$$E_{out} = \frac{E_{in} R2}{R1+R2} = \frac{(20) \times (40)}{20 + 40} = \frac{800}{60} = 13.33 \text{ volts}$$

Thevenin's Theorem—Voltage Division in the Series Circuit (continued)

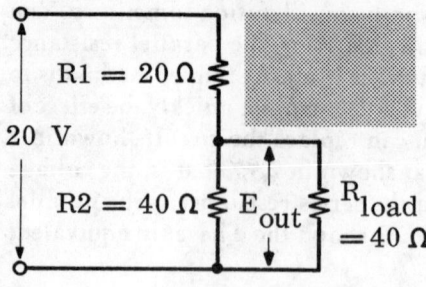

Now, when the load is added, the circuit is changed to that shown at the left. The output voltage is now developed across the parallel combination of R2 and R_{load}, and is equal to 20 ohms. Therefore, the output voltage E_{out} is

$$E_{out} = \frac{(E_{in}) \times R_{parallel}}{R1 + R_{parallel}} = \frac{20 \times 20}{20 + 20} = 10 \text{ volts}$$

You can calculate the *same* thing easier by use of Thevenin's theorem, replacing the voltage source by E_{out} (new E_{in}) without R_{load} and the parallel combination of R1 and R2 as a series resistance as shown below:

$$R_{parallel} = \frac{R1R2}{R1 + R2} = \frac{20 \times 40}{60} = 13.33 \text{ ohms}$$

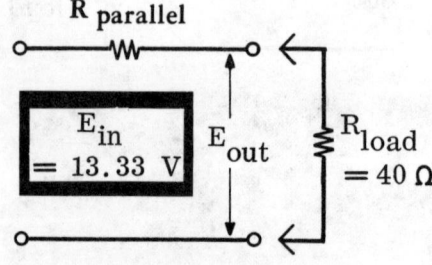

When the resistance load of 40 ohms is added, by using Ohms Law for simple series circuits you can find the output voltage E_{out} when R_{load} is added.

$$E_{out} = \frac{(E_{in}) \times R_{load}}{R_{parallel} + R_{load}} = \frac{13.33 \times 40}{13.33 + 40} = \frac{533.3}{53.33} = 10 \text{ volts!}$$

The answer is the *same!* But you can see that if you wanted to know the effects for many different loads, using Thevenin's theorem will make it much easier to do the calculations. And more importantly, when you are calculating complex circuits, the circuit can often be reduced quickly by applying Thevenin's theorem.

Norton's Theorem—Voltage Division in the Series Circuit (continued)

A similar theorem—Norton's theorem—can be used in situations where a *current* source, rather than a *voltage* source, is more convenient. To use Norton's theorem, the load terminals (across R2) are shorted and the current is calculated.

$$I \text{ (R2 shorted)} = \frac{E_{in}}{R1} = \frac{20}{20} = 1 \text{ ampere}$$

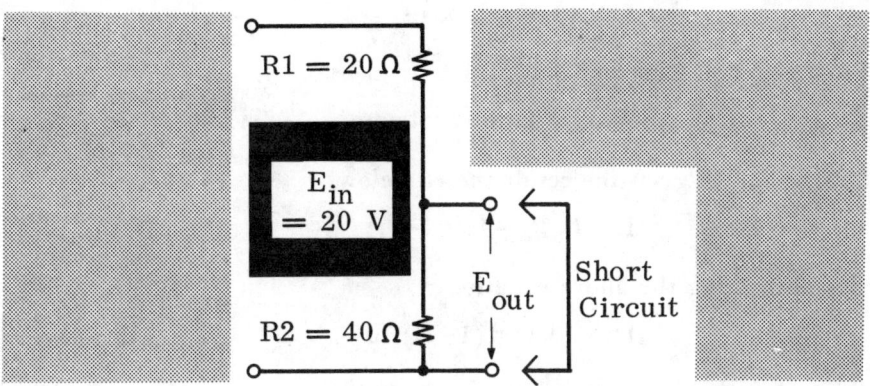

which is the *equivalent current source*. As before, the parallel combination of R1 and R2 is calculated as 13.33 ohms. The Norton equivalent circuit is:

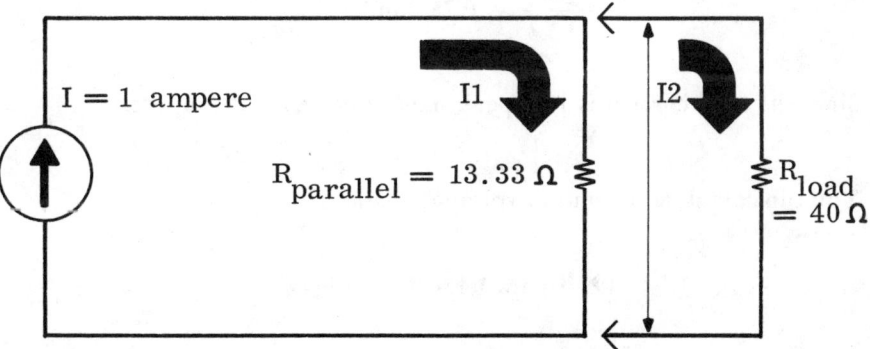

Now without the load, the voltage across E_{out} is 13.33 volts (1 amp and 13.33 ohms), as was calculated using Thevenin's theorem. When the load is added, the current divides into $R_{parallel}$ and R_{load} but the total current is still 1 ampere. In addition, you know that the voltage across $R_{parallel}$ and R_{load} must be the same.

Norton's Theorem—Voltage Division in the Series Circuit (continued)

Therefore, the voltage E_{out} across $R_{parallel}$ is

❶ $E_{out} = I_1 \times R_{parallel} = I_1 \times 13.33$

and also E_{out} is

❷ $E_{out} = I_2 \times R_{load} = I_2 \times 40$

and therefore we can equate 1 and 2

❸ $I_1 \times 13.33 = I_2 \times 40$

From Kirchoff's current law

$$I_{total} = I_1 + I_2 = 1 \text{ amp}$$

Solving for I_2 gives the result shown below

$$I_2 = I_{total} - I_1 = 1 - I_1$$

By substituting the above equation 3

$$I_1 \times 13.33 = (1 - I_1) 40$$

Rearranging terms and solving for I_1 yields

$$13.33 \, I_1 = 40 - 40 \, I_1$$

$$53.33 \, I_1 = 40$$

$$I_1 = \frac{40}{53.3} = 0.75 \text{ amp}$$

Since the total current is 1 ampere, then I_2 must be .25 ampere.

You can calculate the output voltage as either:

$$I_1 \times R_{parallel} \text{ or } I_2 \times R_{load}$$

$$I_2 \times R_{load} = 0.25 \times 40$$

$$= \textit{10 volts!}$$

This is the *same* as was calculated earlier using Thevenin's theorem! Again, Norton's theorem can be useful in solving complex circuits.

Troubleshooting DC Circuits—Basic Concepts

One of the most important things that you must learn to do is to troubleshoot electric circuits. As you proceed in your study of electricity, there will be troubleshooting sections in each volume of *Basic Electricity* that will prepare you for work with various electric circuits. In preparation, you should review the beginning of this volume on what electric circuits are and the meaning of open and short circuits. You then will be able to learn how to troubleshoot dc series, parallel, and series-parallel circuits. If you can troubleshoot these, you then will have the foundation to fix any type of malfunctioning dc circuit.

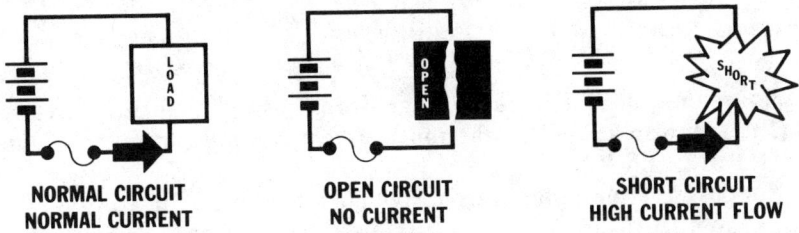

| NORMAL CIRCUIT | OPEN CIRCUIT | SHORT CIRCUIT |
| NORMAL CURRENT | NO CURRENT | HIGH CURRENT FLOW |

As you already know, the two most common problems (faults) in circuits are open circuits and shorts. A short in all or part of the circuit causes excessive current flow. This may blow fuses or burn out components, so a short initially can result in an open circuit.

One of the most important things in acquiring troubleshooting skill is to learn to *use your head* and proceed *logically* through a circuit. Also, to learn to use your senses—to look for loose connections, frayed wires, evidence of overheating, blown fuses, open switches, or plugs not installed. In most cases, you will *see* the problem; you may also *smell* the problem; or even determine the fault by *touch*. If not, then by using your head, that is, using a logical procedure and voltmeters, ohmmeters, and ammeters, you can find the trouble. Remember, a blown fuse means that excessive current was drawn and that indicates a full or partial short circuit. If you replace the fuse and it blows again, you know for sure that you are looking for a short. If you replace a blown fuse and nothing happens, chances are that a short blew out a component of the circuit.

THE BASIS OF TROUBLESHOOTING LIES IN USING A LOGICAL PROCEDURE

By using your senses, your head, and test instruments, you can troubleshoot any electric circuit.

Troubleshooting DC Series Circuits

Suppose you were asked to troubleshoot the series circuit shown:

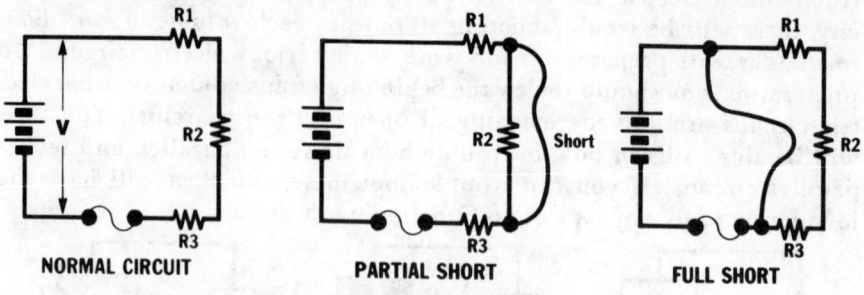

NORMAL CIRCUIT   PARTIAL SHORT   FULL SHORT

You know that the sum of the voltage drops has to equal the source voltage (E). You also know that the total current flows through each component (R1, R2, and R3).

Suppose the symptom is excessive current flow, indicating a full or partial short. If inspection of the circuit shows no frayed or shorted wiring (and you are sure the wiring is correct), the next thing to do would be to determine what potential drop there should be across each resistance. If you calculate these voltages and measure them, the one that is *too low* indicates that this resistance is *low* and should be *replaced*. If the resistances appear to be normal, it is possible the short lies across the entire circuit; measure the resistance across the source (disconnect the power first!), if it is lower than it should be, some connection or misconnection is giving a full short.

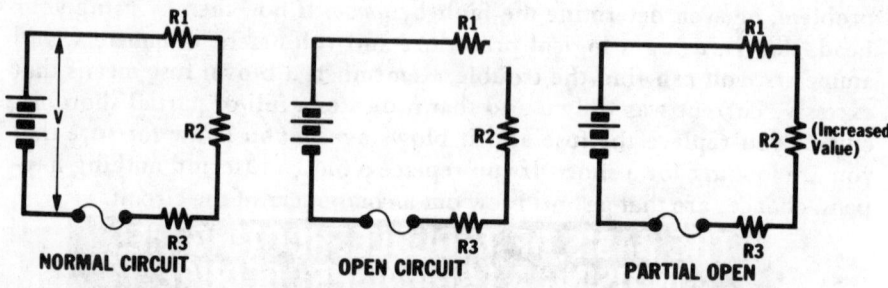

NORMAL CIRCUIT   OPEN CIRCUIT   PARTIAL OPEN

Suppose the symptom is that no current or too little current flows. If no current flows, you should check the voltage source with a voltmeter, and the fuse with an ohmmeter. If these are good, then you must look at the circuit. Inspect the circuit for loose connections or broken wires. After this, proceed to test the voltage drops across each resistance as before; if there is no voltage drop across any resistor, the problem is in the wiring connections. If one voltage is higher than calculated or all the voltage appears across it, then this is the defective component!

Troubleshooting DC Series Circuits (continued)

Example

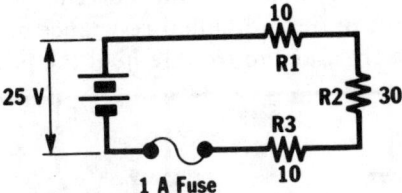

SYMPTOM—*The 1 ampere fuse keeps blowing.*

Inspection shows that the circuit is properly connected and the voltage source is checked with the voltmeter to be 25 volts. By doing the following calculation, you can show that:

The total current should be:

$I_t = 25/R_t$; $R_t = 10 + 30 + 10 = 50$ ohms
$I_t = \frac{25}{50} = 1/2$ A.

The voltages across resistances should be:

$E_{R1} = (1/2) \times (10) = 5$ volts
$E_{R2} = (1/2) \times (30) = 15$ volts
$E_{R3} = (1/2) \times (10) = 5$ volts

Measurement of voltages (or resistances) shows that the voltage R2 is zero. The resistances also measure zero. The current under these conditions is 1.25 amperes and blows the fuse. The corrective or remedial action is to replace resistor R2.

Example

SYMPTOM—*No current flow.*

Inspection shows the circuit to be properly connected. The voltage source is checked with the voltmeter to be 25 volts. Measurement of the voltage across each resistor shows the following:

$E_{R1} = 0$ volts
$E_{R2} = 0$ volts
$E_{R3} = 25$ volts

Thus, it is apparent that R3 is defective.
If the voltages were all measured as zero, then you would have to inspect the wiring since an open lead is indicated.

Troubleshooting DC Parallel Circuits

Troubleshooting parallel circuits is a bit different than for the series circuits shown earlier since for parallel circuits, the voltage across each component is the same and the current through the component depends on the value of the individual resistance of the component.

Suppose you were asked to troubleshoot the parallel circuit shown.

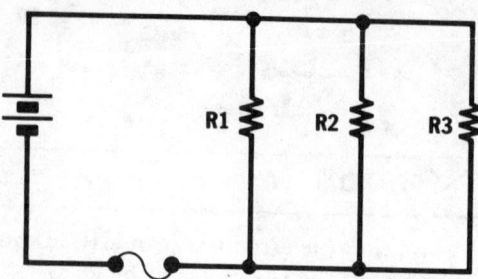

Assume the symptom was excessive current flow, indicating a partial or complete short. Inspection of the circuit shows the wiring to be correct and there are no frayed wires. In the case of parallel circuits, a full short in any component will lead to a full short on the line. It is now necessary to determine *which* component draws excessive current. The quickest thing to do is to check for excessive heat. If this doesn't work, then it is necessary to break the circuit, one component at a time, until the bad (faulty) component is found. You could also put an ammeter in the circuit for each component and find the faulty component that way.

Suppose the symptom was no current flow or less than rated current flow. As with the series circuit, you should check the voltage source and the connections first, to make sure they are correct and that there are no loose or open connections.

Since this is a parallel circuit, an open resistance results only in reduced current flow if there is more than one resistance in the circuit. The presence of the other resistances in parallel makes it difficult to use an ohmeter without disconnecting the individual resistances. Often a parallel circuit will have switches connected to each resistance (or load).

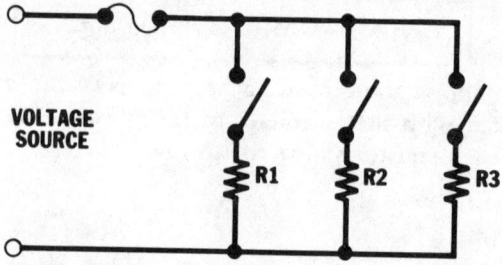

In this case you can isolate the resistances or loads for measurement to determine which one is incorrect (faulty).

Troubleshooting DC Parallel Circuits (continued)

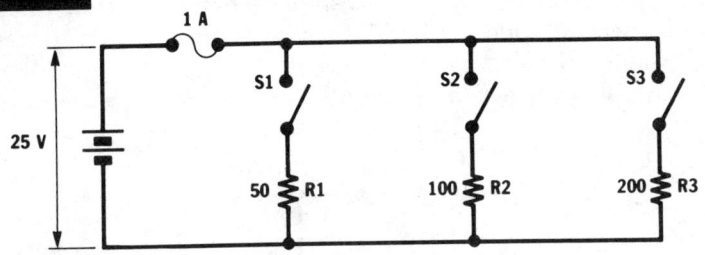

SYMPTOM—*The 1 ampere fuse blows when S2 is closed but not when S1 and/or S3 are closed.*

Inspection shows the circuit is properly connected and does not have miswiring or frayed wires. It should be apparent that R2 should only draw 1/4 amp when connected and the total load should not blow the fuse. The problem, as you have probably figured out, is that R2 has dropped very much in value or is shorted. You could verify this by opening switch S2 and measuring the resistance of R2.

Example

SYMPTOM—*No increase in current flow when S1 is closed.*

Examination of the circuit shows no broken leads or defective or loose connections associated with S1 and R1. You know by Ohm's Law that the current should increase by 1/2 amp when S1 is closed. Also, the parallel resistance with all switches closed should be:

$$\frac{1}{R_t} = \frac{1}{R1} + \frac{1}{R2} + \frac{1}{R3} = \frac{1}{50} + \frac{1}{100} + \frac{1}{200} = 0.035$$

$R_t = 28.57$ ohms

Measurement of the parallel resistance (switches closed and power removed!) shows:

$$R_t = 66.6 \text{ ohms}$$

This doesn't change when S2 is either closed or open, telling you that either the switch is bad or R1 is open. The corrective or remedial action is to replace R1.

Troubleshooting DC Series-Parallel Circuits

Troubleshooting series-parallel circuits is simply an extension of what you already know about series and parallel circuits. You know that you can always break down or rearrange a complex circuit into series and parallel sets of components. You can do this to help in troubleshooting complex circuits.

Suppose you were asked to troubleshoot the complex circuit shown below:

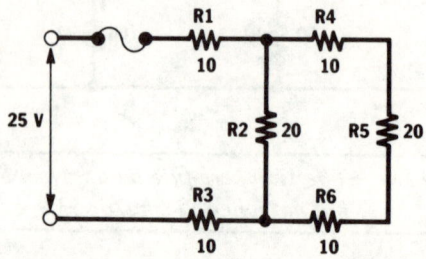

The symptom is that excessive current is drawn. Inspection of the circuit shows frayed insulation on the wires connecting R2 into the circuit. Removal of the short caused by this frayed insulation causes the circuit to act normally.

As you know, the circuit shown above can be connected to an equivalent series circuit by combining R2 in parallel with the series combination R4 + R5 + R6.

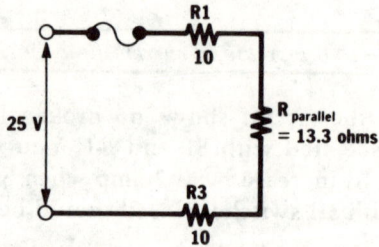

As you can see, this is a series circuit, so you can troubleshoot it as you did for simple series circuits.

TROUBLESHOOTING DC CIRCUITS

Drill in Troubleshooting DC Circuits

1. You have been asked to troubleshoot the series circuit shown below. The symptom is that the fuse blows when the switch is closed. The wiring is correct and there are no obvious short circuits.

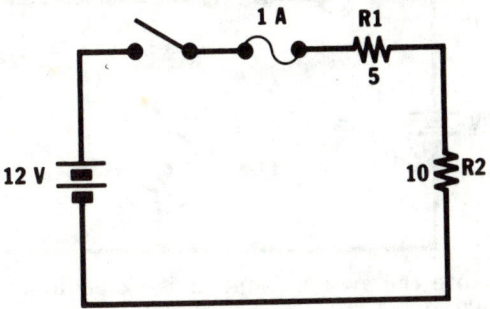

   (a) What current should flow?
   (b) What would you do with an ohmmeter to determine what the problem is?
   (c) What would you do to correct it?

2. In the circuit of question 1, the symptom is that no current flows, the wiring is correct, and there does not appear to be any loose or broken connection. Voltage measurements across the resistance are as follows:

   $ER1 = 0$
   $ER2 = 12 V$

   (a) What is the probable difficulty?
   (b) What measurement could you make to verify your diagnosis?

3. The parallel circuit shown below is part of an automobile electrical system. The symptom is that there are no lights or heat (heater fan) when the switch S2 is closed; however, there is ignition.

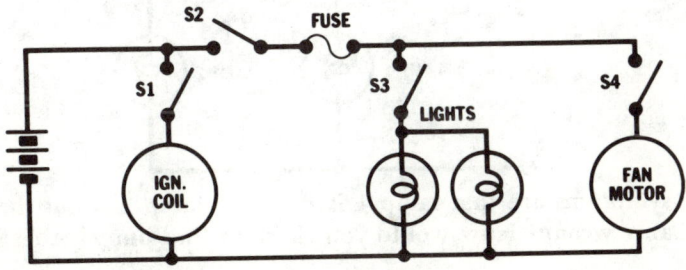

   You find that the fuse is blown. If you replace the fuse and close S2 with S3 and S4 open, the fuse does not blow. Closing S3 makes the lights go on normally. Closing S4 causes the lights to go out and the fan does not run. Inspection shows that the fuse is blown. What do you think the trouble is?

Drill in Troubleshooting DC Circuits (continued)

4. You have the following circuit to operate two 6 V lamps from a 12 volt source.

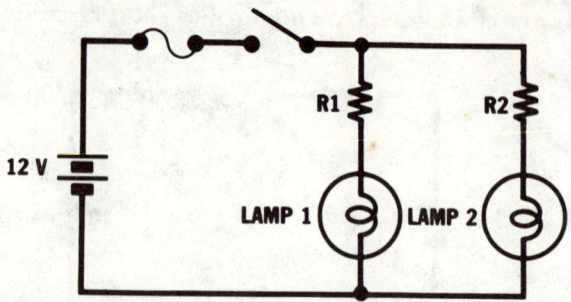

When you close the switch, lamp 2 is exceedingly bright and soon burns out. When you replace the bulb and make voltage measurements, you have the following:

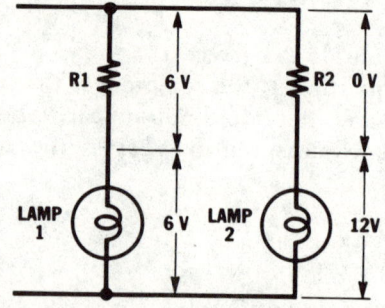

What is wrong? Why?

5. You have been asked to troubleshoot the following complex circuit.

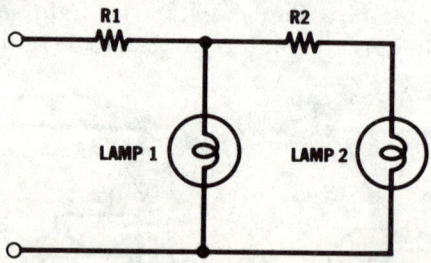

The symptoms are that lamp 1 is dim and lamp 2 is not lit. What is probably wrong? How would you check to find out whether this were so?

REVIEW OF DC ELECTRICITY

General Review of DC Fundamentals

Any combination of a conductor and of a source of electromotive force (emf) which permits free electrons to travel around in a continuous stream from the negative terminal of the voltage source to its positive terminal, and back through the source to the negative terminal again, constitutes an *electric circuit*. As long as this electrical pathway remains *unbroken*, it forms a *closed circuit* and current will flow through it. But if the pathway is *broken* at any point, an *open circuit* results, and no current will flow.

The number of electrons in the electron stream in the closed circuit is dictated by the strength of the emf—voltage—forcing the electrons to move. The magnitude of the electron stream can be restricted and controlled by inserting into the external circuit at any point any kind of *resistor*. The emf is dissipated by the effort of forcing the electron stream through the resistor, and a *voltage drop* takes place across any resistor connected into a circuit.

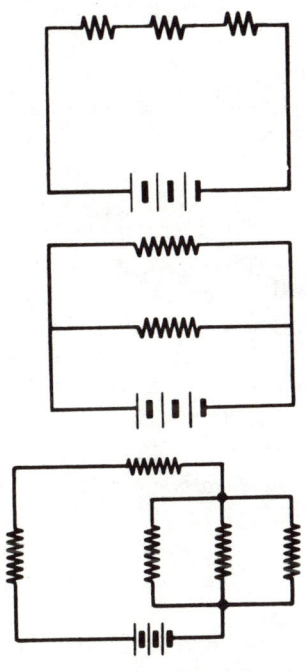

$$\text{CURRENT} = \frac{\text{VOLTAGE}}{\text{RESISTANCE}}$$

1. SERIES CIRCUIT—When two or more resistors are connected *end-to-end* across a voltage source so that the same current flows through all resistors, the circuit is called a *series circuit*.

2. PARALLEL CIRCUIT—When two or more resistors are connected *side-by-side* across a voltage source so that the current divides between them, the circuit is called a *parallel circuit*.

3. COMPLEX CIRCUIT—When a number of resistors are connected into a circuit, some in series and some in parallel, the circuit is called a *series-parallel* or *complex circuit*.

4. OHM'S LAW—There exists a fixed relationship between the voltage, the total resistance of that circuit (or the individual resistance of any resistor connected into it), and the value of current flow through the circuit (or through the individual resistor, as the case may be). This relationship is stated in *Ohm's Law*, which says that *the*

General Review of DC Fundamentals (continued)

$$\text{CURRENT} = \frac{\text{VOLTAGE}}{\text{RESISTANCE}}$$

OR $I = \dfrac{E}{R}$

current flowing in a circuit is directly proportional to the applied voltage, and inversely proportional to the circuit resistance.

In other words, *current flow increases as the voltage is increased, and decreases as the resistance is increased.*

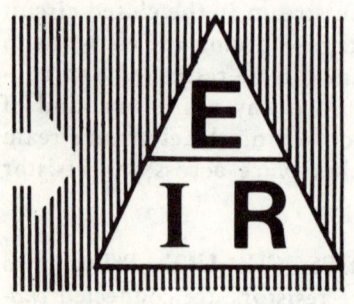

5. THE MAGIC TRIANGLE—You can always find the formula for determining the value of I, E, or R when you know two of the three values, by using the *magic triangle.* Remember the rule: *Put your thumb over the value you don't know—and the formula you want is what's left.*

WATTS = VOLTS × AMPERES

or (in symbols)

$P = EI$

6. POWER—Whenever a voltage causes electrons to move, work is done. The rate at which work is done in moving electrons through a conductor is called *electric power.*

Electric power is represented by "P" and is measured in watts, "W"; 1 watt is defined as *the rate at which work is being done in a circuit in which a current of 1 ampere is flowing when the emf applied is 1 volt.*

The *power formula* states that the power consumed in a resistor is determined by the voltage across it multiplied by the current flowing through it.

$P = I^2 R$ and $P = \dfrac{E^2}{R}$

Power can be determined when circuit resistance is known, but either current or voltage is not known, by substituting Ohm's Law for the unknown factor in the equation, $P = EI$. The rewritten equations are:

$$P = I^2 R \quad \text{and} \quad P = \frac{E^2}{R}$$

Learning Objectives—Next Volume

Overview—You have learned about dc circuits. In the next volume—Volume 3—you will learn about alternating current or ac circuits. You will see that what you know about dc circuits can be applied to ac circuits.

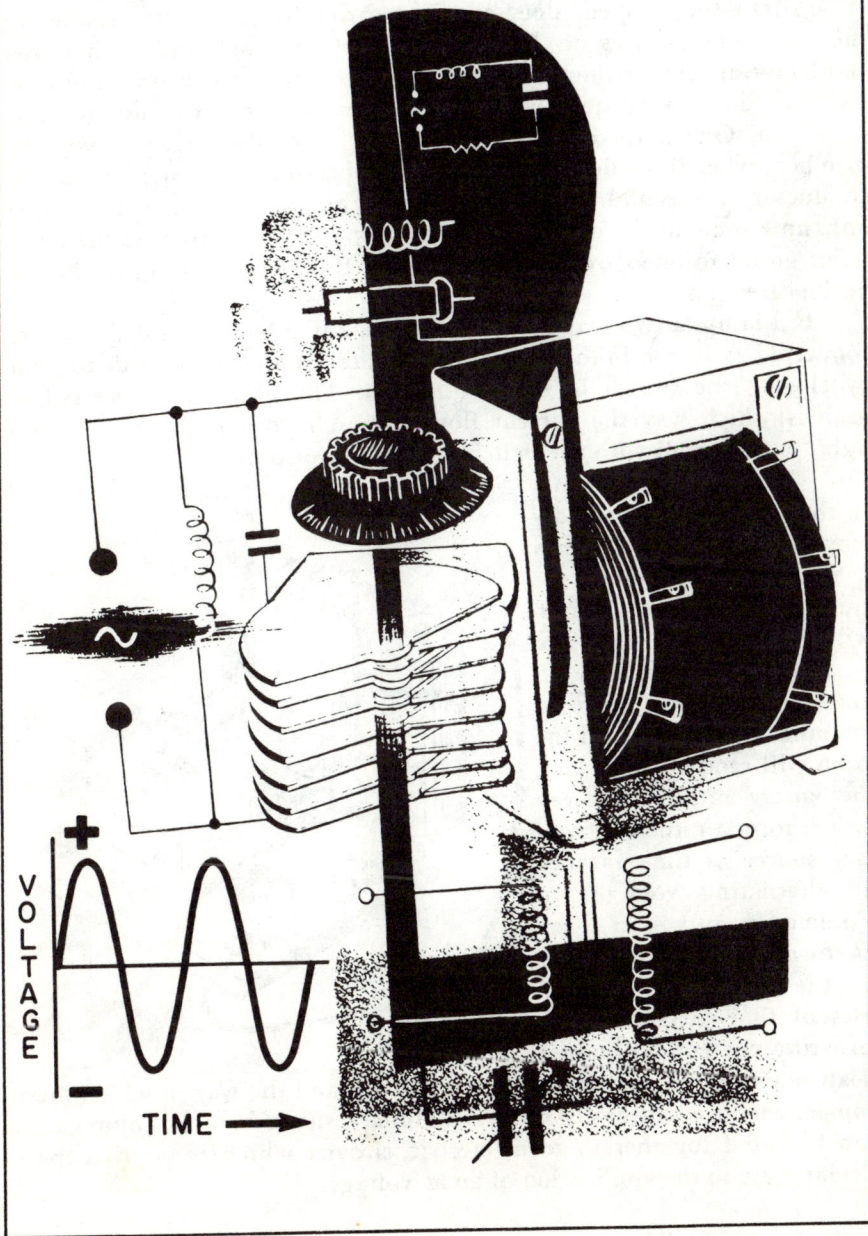

Alternating Current

The circuits you have been studying so far have all been direct current (dc) ones. In practice, however, very little direct current is used to supply electric light or power; and, furthermore, nearly all electronic circuits make at least as much use of what is called *alternating current* (ac) as they do of dc. So you now must start to learn what this ac is and how it behaves.

Alternating current does not flow through a conductor always in the same direction, as dc does. Instead, it flows back and forth in the conductor at regular intervals, continually reversing its direction of flow and can do so very quickly. It is measured in amperes, just as dc is measured. One ampere of current is said to be flowing, you will remember, when 1 coulomb of electrons is passing a given point in the conductor in 1 second. This definition also applies when ac is flowing—only now some of the electrons during that 1-second flow past the given point going in one direction, and the rest flow past it going in the opposite direction.

If a lamp, a two-way (DPDT) switch, and a battery of dry cells are connected as shown in the diagram below, the lamp can be made to light by closing the switch in either direction. The position of the switch decides which way the current flows through the lamp, but the lamp lights equally well with the switch in either position.

If the switch is repeatedly moved very quickly from one position to the other, the current through the lamp will *alternate* or *vary with time*—that is, first flow in one direction and then in the other direction—and the lamp will remain lit. In fact, the battery and the two-way switch form a rather elementary source of time varying or alternating voltage. In Volumes 3 and 4 of *Basic Electricity*, you will be studying the nature, behavior and uses of time-varying or alternating current. Volume 3 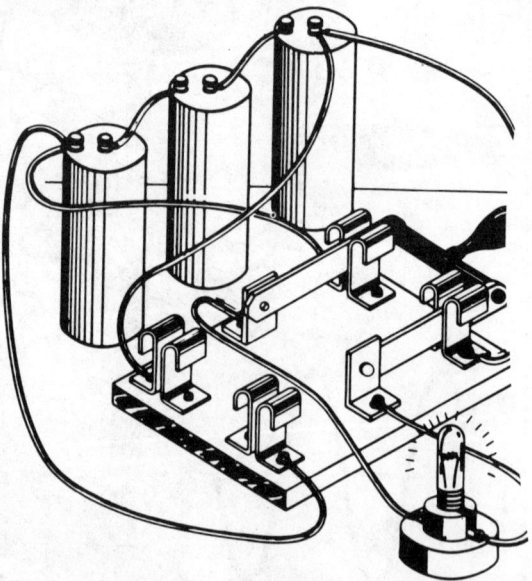 deals mainly with ac terms and components, and the way in which these components fundamentally behave. Volume 4 shows how the components can be fitted together to form electric circuits which respond in particular ways to the application of an ac voltage.

INTRODUCTION TO ALTERNATING CURRENT 2-149

Alternating Current (continued)

Also in Volume 3, you will meet for the first time two components —the *inductor* and the *capacitor*—which are frequently used to control direct as well as alternating current and voltage. The resistors with which you have been working so far have all acted in such a way as to restrict the flow of current *directly*. In other words, the bigger the resistor you put in, the more you restrict current flow. The inductor and the capacitor, on the other hand, act to control current and voltage in rather different ways, and you will see that what they do depends on how often the current is reversed. These three components—the resistor, the inductor, and the capacitor—are basic elements of all electric and electronic circuits.

Answers to Drill Questions

Page 2-16

1. (a) $E = IR = 2 \times 10 = 20$ volts

    (b) $R = \dfrac{E}{I} = \dfrac{30}{1.5} = 20$ ohms (Ω)

    (c) $I = \dfrac{E}{R} = \dfrac{10}{15} = 0.667$ ampere

    (d) $I = \dfrac{E}{R} = \dfrac{300}{1,000,000} = 0.0003$ ampere $= 300\ \mu A$

    (e) $E = IR = 2,500,000 \times 0.000002 = 5$ volts

    (f) $R = \dfrac{E}{I} = \dfrac{495}{0.0015} = 330,000 = 330$ K

2. $R = \dfrac{E}{I} = \dfrac{12}{4} = 3\Omega$ is the lamp resistance.

3. To make rated current flow, the voltage required is
    $E = IR = 1.5 \times 24 = 36$ volts

4. The element resistance must be
    $R = \dfrac{E}{I} = \dfrac{240}{2.5} = 96$ ohms

5. $I = \dfrac{E}{R} = \dfrac{1.36\text{ V}}{68\text{ K}} = \dfrac{1.36}{68,000} = 0.00002$ ampere $= 0.02$ mA $= 20\ \mu A$

6. $R = \dfrac{E}{I} = \dfrac{10\text{ V}}{5\text{ mA}} = \dfrac{10}{0.005} = 2,000$ ohms (Ω) $= 2$ K

Answers to Drill Questions

Page 2-34

1. In a series circuit, the total resistance is the sum of the resistances.

    R_t = R1 + R2 + R3 ...

        = 220 ohms + 680 ohms + 1,000 ohms

        = 1,900 ohms = 1.9 K

2. R_t = R1 + R2 + R3 + R4

    Therefore,   67 = 10 + 15 + 27 + R4

    and   R4 = 67 − 10 − 15 − 27

                    = 15 ohms

INDEX TO VOL. 2

(Note: A cumulative index covering all five volumes in this series will be found at the end of Volume 5.)

Alternating current, 2-148 to 2-149

Circuits, 2-1 to 2-10
  ac and dc, 2-2 to 2-3
  nature of, 2-4 to 2-5
  open, 2-47 to 2-49
  parallel, 2-59 to 2-85
  series, 2-32 to 2-58
  series-parallel, 2-91 to 2-111
  short, 2-50 to 2-52
  troubleshooting, 2-137 to 2-146
Complex circuits, 2-121 to 2-122
Connections, 2-8

Electric power, 2-112 to 2-132
  complex circuits, 2-121 to 2-122
  fuses, 2-117 to 2-118
  power formula, 2-113 to 2-114
  parallel circuits, 2-120
  ratings, 2-115 to 2-116
  series circuits, 2-119
  work, 2-112
Experiments/Applications
  Electric Power, 2-125 to 2-132
  Ohm's Law, 2-19 to 2-21
  Parallel Circuits, 2-88 to 2-90
  Parallel DC Circuits, 2-66 to 2-72
  Series DC Circuits, 2-47 to 2-57
  Series-Parallel Circuits, 2-108

Fuses, 2-117 to 2-118

Kirchhoff's Laws, 2-57, 2-70 to 2-75

Loads, 2-6

Norton's Theorem, 2-133 to 2-136

Ohm's Law, 2-11 to 2-21, 2-82
Open circuits, 2-47 to 2-49

Parallel circuits, 2-59 to 2-85
  current in, 2-61 to 2-63, 2-67
  Ohm's Law, 2-82
  power, 2-120
  resistance in, 2-64 to 2-65, 2-68 to 2-69

  solving unknowns, 2-83 to 2-85
  troubleshooting, 2-140 to 2-141
  voltage in, 2-60 to 2-66
Power formula, 2-113 to 2-114
Power ratings, 2-115 to 2-116

Resistance, 2-27 to 2-31
Resistors, 2-22 to 2-26
  unequal, 2-76 to 2-79
  variable, 2-42 to 2-44
Reviews
  DC, 2-145 to 2-146
  Electric Circuits, 2-9 to 2-10
  Electric Power, 2-123
  Ohm's Law, 2-45
  Parallel Circuits, 2-80, 2-86
  Resistance, 2-30 to 2-31
  Series-Parallel Circuits, 2-108

Series circuits, 2-32 to 2-58
  current in, 2-35, 2-56
  power, 2-119
  resistance in, 2-27 to 2-31, 2-53 to 2-55
  troubleshooting, 2-138 to 2-139
  voltage in, 2-37, 2-41 to 2-45
Series-parallel circuits, 2-91 to 2-111
  bridge circuits, 2-98 to 2-100
  Ohm's Law, 2-101 to 2-106
  resistors, 2-92 to 2-97
  troubleshooting, 2-142
Short circuits, 2-50 to 2-52
Switches, 2-2 to 2-7

Thevenin's Theorem, 2-133 to 2-136
Troubleshooting dc circuits, 2-137 to 2-146
  concepts, 2-137
  drill, 2-143 to 2-144
  parallel circuits, 2-140 to 2-141
  series circuits, 2-138 to 2-139
  series-parallel circuits, 2-142

Unequal resistors, 2-76 to 2-79

Variable resistors, 2-42 to 2-44

Work, 2-112